Deschutes
Public Library

THE EDGE OF MEMORY

Also available in the Bloomsbury Sigma series:

THE EDGE OF MEMORY

ANCIENT STORIES, ORAL TRADITION AND THE POST-GLACIAL WORLD

Patrick Nunn

BLOOMSBURY SIGMA

LONDON · OXFORD · NEW YORK · NEW DELHI · SYDNEY

BLOOMSBURY SIGMA
Bloomsbury Publishing Plc
50 Bedford Square, London, WC1B 3DP, UK

BLOOMSBURY, BLOOMSBURY SIGMA and the Bloomsbury Sigma logo are
trademarks of Bloomsbury Publishing Plc

First published in 2018

Photo credits (t = top, b = bottom, l = left, r = right, c = centre)
Colour section: P. 1: Simon Albert (t); Auckland Art Gallery Toi o Tāmaki, transferred
from Taranaki Museum, 1994 (b). P. 2: Art Collection 2/Alamy Stock Photo (t); [PD-
old-100, PD-US] (b); P. 3: Keith Morris News/Alamy Stock Photo (t); Herbert Felton/
Stringer/Getty Images (bl); Hulton Archive/Stringer/Getty images (br). P. 4: Patrick
Nunn. P. 5: Dan McShane (t); Aditya Prakashan and S. R. Rao (cl); [PD-old-100] (b).
P. 6: *Ngurriny – The Fly Dreaming* © Jane Gorden/Copyright Agency, 2018 (t); Patrick
Nunn (bl, br). P. 7: Paul Biris/Getty Images (t); Fortune Dragon Group (c); Creative
Commons Attribution-Share Alike 2.0 Generic (b). P. 8: Lee Scott-Virtue, Jack
Pettigrew and Dean Goodgame (tl); Peter Schouten (tr); Robert Gunn / Margaret
Katherine (b); Peter Trusler © Australian Postal Corporation 2008 (b, inset).

A catalogue record for this book is available from the British Library

Library of Congress Cataloguing-in-Publication data has been applied for

ISBN: HB: 978-1-4729-4328-6; TPB: 978-1-4729-4326-2;
eBook: 978-1-4729-4327-9

2 4 6 8 10 9 7 5 3 1

Illustrations by Patrick Nunn

Typeset by Deanta Global Publishing Services, Chennai, India
Printed and bound in Great Britain by CPI Group (UK) Ltd, Croydon CR0 4YY

Bloomsbury Sigma, Book Thirty-nine

To find out more about our authors and books visit www.bloomsbury.com
and sign up for our newsletters

For HB and Mz

The author would like to acknowledge with respect and gratitude the originators and countless communicators of the remarkable stories covered in this book. He also thanks the traditional custodians of the land in the part of Australia where this book was written – the Gubbi Gubbi (Kabi Kabi) people – and their elders past and present whose spirits imbue the place with life and meaning.

Contents

CHAPTER ONE

Recalling the Past

It was the year 1853, a balmy June afternoon in western Oregon, USA, and John W. Hillman, late of Albany, New York, was lost. Riding a mule, Hillman led his party of seven men up 'a long, sloping mountain' so they could try and figure out where they were. It was a good thing that the mule was not blind, Hillman reflected 50 years later, because otherwise he might have been the first New Yorker to drown in the lake that appeared unexpectedly almost 600m (1,968ft) below him when he reached the mountaintop.[1] The lake was Crater Lake and the Hillman party is generally credited with providing the earliest written account of it. But they did not 'discover' it, for its existence was well known to the indigenous people of the district, principally the Klamath Indian tribes.

As gold seekers and settlers poured into this region over the next few decades, supported by government representatives, so the apparently strange attitude of the Klamath towards Crater Lake was increasingly remarked upon. For example, in 1886 a US Geological Survey reconnaissance group included two Klamath guides who were familiar with the entire area around the lake, 'neither of whom had dared travel to Crater Lake before'.[2] Around the same time, William Steel, intent on setting up a national park centred on Crater Lake, noted that the Klamath people he engaged refused to look at the lake at any time during his survey, instead 'making all sorts of mysterious signs and staring directly at the ground'.[3] Enquiries revealed that to the Klamath, Crater Lake was a sacred place, one greatly respected and to be avoided by all except their shamans, who went there only when needing divine guidance.

What lies behind such behaviour and is it unique? In the case of the Klamath, the avoidance and respect behaviours are reportedly rooted in a memory of a time before Crater Lake existed, when the entire area was covered by a massive volcano, belatedly named Mt Mazama. One of some 18 active volcanic mountains forming the Cascade Range, from Lassen Peak in northern California to Silverthrone Caldera in western British Columbia, Crater Lake – and its long-dead ancestor Mt Mazama – formed along a line parallel to a giant 'crack' in the surface of the Earth's crust off this part of North America. The crack marks the place where the crust underlying the vast Pacific Ocean is being thrust eastwards beneath the older continental crust of North America – reluctantly, it seems, for the movement of one piece of crust (or 'plate') under an adjoining piece progresses through stick-slip motion. Most of the time the two plates are stuck together, locked in place, but all the time they continue to be pushed more and more towards one another ... until at last there is a slip, pressure is abruptly released (causing an earthquake) and the plates slip past one another.

This explains why this part of the western seaboard of North America is wracked by earthquakes compared to other parts of the continent. Sometimes the earthquakes occur on the ocean floor, abruptly displacing huge volumes of water and causing tsunamis. One of these occurred on 26 January 1700 off the central Washington coast,[4] and sent a large wave across the North American shore – yet the only written records from this time come from Japan, which was also reached by this distantly generated tsunami.[5]

The slippage that causes earthquakes is only one aspect of the convergence of the crustal plates that occurs here. Another is the eruptions that have – over the past 40 million years – built the line of Cascadia volcanoes some 150–250km (93–155 miles) from the continent's western edge. Actually more than 2,000 in number, these volcanoes trace the line deep below the ground surface where the downthrust Pacific Plate has become so hot that it melts, perhaps some 100km (62 miles) down where temperatures are more than 650°C (1,200°F). The melting produces liquid rock (or magma), which then – finding itself surrounded by cooler solid rocks – tries to force its way upwards through the overlying solid crust. Much of the time, probably even most of the time, it does not succeed in reaching the ground surface because most of the fissures it encounters do not extend far enough upwards, so the magma pools below the surface, where it gradually cools. But for the enterprising magma that does find a clear path to the ground surface, a more spectacular destiny is guaranteed.

Where magma rushing upwards from below through a constrained fissure finally comes out to the ground surface, the rapid cooling it experiences often results in explosive eruptions and – less spectacularly – the successive deposition of lavas that gradually build a volcano above the mouth of the subterranean fissure. As these volcanoes become larger, and perhaps as the supply of magma from beneath becomes progressively less, so it often becomes more difficult for eruptions to occur at the summit of the volcano. In such cases, flank (or parasitic) cones may develop at lower levels. Occasionally, the magma supply dries up altogether and a volcano is declared extinct, or at least dormant. Thereafter it is doomed to slowly degrade, often becoming worn down so as to appear almost indistinguishable from the surrounding landscape. This certainly seems to be what is happening with the Canadian Cascade volcanoes. Silverthrone has not erupted for an estimated 100,000 years and is probably extinct. But then most volcanoes go through periods of dormancy, often lasting centuries, before springing back into activity. Mt St Helens is one of these volcanoes – its present period of activity, the centrepiece of which was the eruption on 18 May 1980, followed two periods of dormancy lasting more than a century. Enough to lull anyone into a false sense of security, you might suppose.

You see the fateful consequences of this mindset in many places, not just the north-west United States, of course. It occurs where new people arriving to live in a potentially hazardous area may be unaware of the danger, while longer-term residents are more wary. This contrast is amplified when the new arrivals are literate and the long-term occupants are non-literate, since literacy often confers arrogance and an uncritical faith in the superiority of the written over the spoken word. Around Crater Lake, settlers in the late nineteenth century were generally uninterested in the cautionary stories and behaviours of the Klamath, but more than that, Western science has been slow to consider the possibility that such folk memories may have been based on observations of a geological phenomenon – in this case a massive volcano-destroying eruption – empirical evidence of which has been pieced together only within the last few decades.

The Klamath story holds that within Mt Mazama resided a god – the Chief of the Below World – who desired to wed the beautiful Loha, a human living in a nearby village, but she resisted his blandishments.[6] Enraged, the Below-World Chief began to rain red-hot rocks and burning ash down on Loha's people, but they were saved through the intervention of the Above-World Chief, who eventually caused the mountain to collapse inwards on his underworld counterpart. The hollow that formed filled with water and thus Crater Lake came into being. In 1865, a Klamath elder explained how these events had shaped his people's attitude towards the area:

> Now you understand why my people never visit the lake. Down through the ages we have heard this story. From father to son has come the warning, 'Look not upon the place. Look not upon the place, for it means death or everlasting sorrow.'[7]

The Klamath stories are interpreted as recalling a time before Crater Lake existed, when Mt Mazama towered over the landscape of this part of Oregon. They recall its explosive self-destruction, the ensuing collapse and the formation of the lake – a process well known to geologists studying the life cycles of such massive volcanoes. To understand what happens, consider that large volcanoes like Mt Mazama are often fed from a shallow underground chamber that periodically fills with magma. In this case, the voluminous eruption that saw this volcano blow itself to pieces was fed from a magma chamber about 5km (3 miles) deep. So much magma came out that the magma chamber was suddenly emptied – a massive void under the ground remained and, unable to support the weight of the overlying volcano, collapsed. Today, where Mt Mazama once stood, the eponymous crater lake occupies a landform called a caldera that formed above the collapsed magma chamber.

What is really interesting here is that the radiometric dating of rocks belonging to the former Mt Mazama show that its self-destruction occurred some 7,600 years ago.[8] We are left to wonder how the story of the formation of Crater Lake could have endured so long among people whose only method of recalling their history, at least until about 150 years ago, was through word of mouth ... parent to child, parent to child across almost countless generations. It was a story of history, of course, but also one with a warning: keep away from this place, it is dangerous; terrible things happened here, involving beings who are more powerful than us mortals, beings whom we suspect still dwell here.

It is likely that the memory of Mt Mazama was kept alive for such an extraordinarily long time precisely because it was considered to contain such practical advice relating to the survival of the Klamath people. But the astonishing thing about the story is its longevity. How could an eyewitness account of events 7,600 years ago filter down to us today, almost entirely through intergenerational oral transmission? And what does this mean for our understanding of humankind? Somewhat against the grain of the last 100 years or so of scientific inference, this book argues that stories of this kind do exist in many of the world's cultures, but that their apparent longevity has led to their dismissal by generations of scientific commentators who have been unable to overcome their knee-jerk scepticism about the implied time depth of oral tradition. Perhaps it is time to look less circumspectly, with a more open mind, at traditions of possible great longevity elsewhere.

History celebrates memorable events and shuns the mundane. It is in our nature to do this – to wish to have our minds stimulated and our imaginations unleashed – rather than to be compelled to focus on predictable everyday happenings. In each decade of adulthood we are prone to look back at our lives, illuminating and calibrating our personal histories with light-bulb pops of uncommon events. Now imagine the personal stretched to the communal, the journey through time of a particular people living in a particular place and trying to make sense of their history without the aid of literacy.

A fine example comes from the islands of Pukapuka in the Cook Islands group of the South Pacific. A jewel in a turquoise sea, inhabited for a millennium or more by a resilient people following a way of life optimally adapted to a comparatively low, remote and resource-constrained tropical location. Of course, we can paint lives in contrasting ways, and probably most modern Pukapukans would consider themselves blessed, fortunate to be living in a bountiful environment immune from the pressures that many urban dwellers routinely experience. But it has not always been so on Pukapuka, the traditional (largely oral) history of which is in two parts, separated by what the islanders recall as *te mate wolo* – the great death. As near as we can tell, it happened at night sometime in the early decades of the seventeenth century, when a giant wave – almost certainly a tsunami generated by an ocean-floor earthquake off the western margin of the Americas, perhaps Chile – smashed into the island. The oral histories remember that the 'waters raged on the reefs, the sea was constantly rising, the tree tops were bending low'.[9] Everything was destroyed; no houses and no food gardens remained in the aftermath. The traditions state that just two women and 17 men 'with remnants of their families' survived to re-establish human society on Pukapuka.

Another example comes from Sirente in central Italy, during the late Roman period around 1,650 years ago. This marked a time when Christianity was slowly gaining a foothold in the region, displacing supposedly pagan cults like that which embraced the Dionysian ritual practices of ecstatic dancing and licentious behaviour following the annual grape harvest. The story from Sirente, dating from about the year

AD 412, describes the start of a ritual in a mountainside temple around sunset:

> *Dances and songs were practised, along with clever and ribald witticisms. Dishevelled Bacchanalians languished under the effect of wine. Hairy satyrs, human at the top except for their goat's ears and legs, danced with snow-white nymphs and the corpulent and sunburnt Silèni. The nymph Sìcina was by far the most beautiful, shameless and bold. Around her, the orgy would become intense, almost violent.* [10]

But then an 'uproar hit the mountain', trees were split apart and a 'sudden and intense heat overwhelmed the people':

> *All of a sudden ... a new star, never seen before, bigger than all the other ones, came nearer and nearer, appeared and [then] disappeared behind the top of the eastern mountains. People's eyes looked at the strange light growing bigger and bigger ... an irresistible, dazzling light pervaded the sky.*

The 'star' was a meteorite, glowing as it passed over the Dionysians to crash to Earth in the neighbouring valley with a massive thump that shook the mountains. While the force of the explosion caused the meteorite to vaporise, the crater its impact made remains clearly visible, and samples from the compacted ground surface have allowed the time of the collision to be estimated. There was no science available at the time to allow the revellers to make sense of what had happened, so they interpreted it as a sign to embrace Christianity, something they did with immediate, uncompromising and enduring fervour. [11]

For most of the time since the fall of this meteorite, the story of its impact on local beliefs was passed on orally from one generation to the next, and it was not written down until the 1890s. It is easy to believe that to discourage recidivism the people of the Sirente wished to keep alive the lesson their ancestors had been given, but what is more remarkable is the inescapable fact that the story was passed down by word of mouth across so many generations, perhaps 60–70, without its essence being lost.

This is remarkable by modern standards, of course, for most of us could readily name instances of imperfect recollection in our daily lives. But then most of us have grown up reading, dependent on the written word – and its visual extensions – for knowledge of just about everything we know at the flick of a page or the press of a button. We have become dependent on writing and reading, and in the process have invariably convinced ourselves that the written word is superior – must be superior – to the spoken word. But it was not always so.

Think about why we communicate, why we speak, why we write. In today's modern, globalised world, the reasons are many – to inform, to counsel, to negotiate, to express our feelings. In the past, when human survival required generally less information and involved fewer choices, the same four reasons would also have applied, but would have been far more constrained. Take food choices as an example. In the course of a single day, many of today's urban dwellers routinely have to make food choices – in stores, markets and restaurants – that their ancestors 1,000 years earlier could not have imagined. Food in such times was a mere part of life, inadequate for many, and rarely a subject involving choice. Today in many richer countries, food is not just a matter of choice but is often emblematic, its methods of preparation and serving having almost countless permutations – all of which, of course, need words to explain, words that might be difficult for each of us to remember accurately were it not for writing.

Writing therefore helps us to manage knowledge when its volume threatens to become too great for us to remember otherwise. When faced with a need to undertake complex tasks, the capacity of our memory often proves insufficient. In non-literate societies the amount of information at an individual's disposal is likewise finite, so they are not generally able to undertake complex tasks (like brain surgery, satellite building or establishing zoos), which people inhabiting a literate world have come to do. Yet writing did not evolve because people yearned to undertake complex tasks – it probably came about incrementally, in the name of pragmatism, because it was needed to help societies of increasing complexity manage themselves. Writing helps when things need to be counted, checked, weighed, then communicated accurately to people beyond earshot. Fortuitously, writing was also found to have the power to allow one individual to become better informed than others from whom particular knowledge could be kept. It became a weapon that was often wielded by a small elite in order to repress a majority. The subsequent invention of printing presses marked the beginning of the end of this, and written knowledge of almost every kind is potentially at everyone's fingertips today.

The focus of this book is not on writing but on its predecessor – speech – and the way this was used to transmit knowledge from one generation to the next. In today's literate societies, speech is generally reserved for relatively simple communication, whereas writing (and its scripted visual counterparts) is the method by which our most complex thinking is commonly shared. This situation underlines the point that speech alone is not generally able to adequately communicate all the complexities of the world as we know them today. Yet speech in non-literate societies may also not have been adequate to capture all of life's complexities, which is why so many such societies evolved behaviours – perhaps like the bacchanalia of Sirente – that may have been difficult even for their contemporary practitioners to adequately explain. This is because speech was the principal means by which non-literate societies communicated between the older and the younger generations – passing on knowledge through the ages – so it had to be optimally configured for this purpose.

Modern humans (*Homo sapiens*) first appeared on Earth about 200,000 years ago in tropical Africa.[12] From there, as some of their hominid ancestors had done, our species dispersed to other parts of the world.[13] Being smarter than these ancestral hominids, *Homo sapiens* asked questions their ancestors may not have thought to ask. How might we slaughter these animals many times our size? Can we eat this? How can we fashion tools to enable this? Instead of going around this large body of water to reach the other side, is there a way we can cross it? Groups of cooperating modern humans may have crossed the Red Sea 130,000 years ago, hopping from one island to another through the Farasan group to reach the coast of Asia. Arriving there, in what is probably the earliest example of human impact on natural resources, they gorged themselves on nearshore seafoods, even eating one particular shallow-water clam species almost to extinction.[14]

We can envisage situations where rudimentary speech evolved among these groups of humans to enable the crossing of small water gaps and the exploitation of unfamiliar food sources, but it was later, perhaps 60,000–70,000 years ago, that language became requisite. By then, *Homo sapiens* had followed the coast of southern Asia[15] into what are now the island coasts of Indonesia – places like the south coast of Borneo and the north coast of Java – that were at that time, when the sea level was 80m (260ft) lower than it is today, contiguous with the rest of dry-land Asia. People found themselves in a situation where they wished to cross the ocean to reach the land they were convinced existed over the horizon.[16] To achieve this they needed watercraft, and irrespective of whether these were simple bamboo rafts lashed together with vines or something altogether more sophisticated, it was not possible for one person to do everything alone. Cooperation was needed, direction was required, so effective communication using language is regarded as having been essential by this stage in modern human evolution.[17]

As discussed in more detail in Chapter 4, the level of the ocean surface (sea level) changed by 100m (328ft) or more every 100,000 years or so within the past few million years. These great swings of sea level repeatedly altered the coastal geographies of most places, but would have been most noticeable in regions of what are now islands – like South-east Asia – yet were once contiguous continental land masses. Those ancestors of ours, naked with sun-browned skins,[18] who reached the south-east corner of the Asian continent some 70,000 years ago when the sea level was 60m (200ft) lower than it is today, found that there was no more dry land by which they might extend their range southeastwards. They had reached the edge of a deep-water passage marking the Wallace Line, the name given by zoogeographers to the faunal boundary between the Asian and Australian continents, which no land mammals before humans had been able to cross.[19]

The cross-ocean journeys that took place 60,000–70,000 years ago in island South-east Asia involved the successful traversing of distances as great as 70km (43 miles), and resulted in the first human arrivals in Australia perhaps 65,000 years ago.[20] The successful and sustained colonisation of Australia at this time was almost certainly aided by the ability of the first Australians to communicate through speech. Life in Australia was hard, ultimately proving more challenging than it may have been along the tropical coasts of Asia. Not only was much of Australia uncommonly dry, but it was also inhabited by animals and plants with which the colonists were unfamiliar. It is likely that language became key to adaptation and survival. Knowledge about water sources, and about where to find food, and how to capture and consume it, probably all became part of a lore that was intentionally passed on from one generation to the next to ensure a tribe's survival. This is certainly what ethnographic information from Australian Aboriginal groups collected tens of thousands of years later suggests.[21]

Elsewhere in the world, the evolution of language also became key to the ability of *Homo sapiens* to survive in environments outside those in which it had originated. Clothing provides a good example of this ability. In many tropical climates, clothing to protect our bodies from the weather is not necessary to survival – we can be naked – but the movement of modern humans into cooler, higher-latitude environments required us to find ways of enduring their colder, even freezing, conditions. Our ancestors therefore began using animal skins, learning how to treat them so that they would last longer. Several human groups experimented with clothing made from tree bark, a process that fortuitously led to the invention of paper – the essential partner of language on the journey to literacy – in several Asian and European cultures. All these comparatively complex processes required cooperation between humans that could only – it is thought – have come about after language of sufficient complexity had evolved.[22]

Plant domestication – essential for feeding high-density populations – also required people to communicate through language. Which plants were suited to human consumption, how they might be cultivated, nurtured, harvested and prepared, were all questions that required the use of memory and communal knowledge sharing, underpinned by sufficient language abilities. In this way, language facilitated the rapid evolution of human societies that occurred in several parts of the world within the past few millennia. The transformation was from lifestyles based on hunting and gathering, which often required people to be nomadic, through the cultivation of crops and domestication of animals that allowed sedentary self-sufficient communities to be established, to the development of urban centres that housed hierarchical societies in which most residents were not directly involved in food production. They became the first consumers to be removed from the front line of production: the earliest merchants, food processors like bakers and butchers, and of course the priests and the soldiers responsible for others' welfare.

The earliest urban centres in the world were in those places – often close to fertile, well-watered river floodplains – where large numbers of people could be fed from cultivated crops and domesticated animals. There were some such places in southern China, where rice agriculture began perhaps 7,000 years ago in the lower valleys of the rivers Huanghe and Yangtze.[23] Others developed in Mesopotamia and depended on a range of agro-pastoral activities, such as dry-farmed wheat, barley and lentils supplemented by sheep and goat husbandry.[24] Complex, comparatively densely populated societies of this kind involved organisation and required management to endure, both of which in turn necessitated language of appropriate complexity. It is no coincidence that writing evolved more rapidly in such societies than those elsewhere.

In other parts of the world – in Australia, parts of Africa and the Americas, for example – opportunities for plant and animal domestication were fewer and the pressure for this less exigent, so people continued to feed themselves largely through hunting and gathering. Language in such situations may have developed as the need arose for new words and concepts; it therefore became different in both vocabulary size and descriptive focus from the way that language evolved in cities. What is of considerable interest to this book are the cultural dimensions of language evolution – the development of words and concepts that describe worlds of the mind.[25] The links between imagination and myth, especially as they relate to human memories of calamitous events, are of especial interest.

In March 1847, Mt St Helens – one of the volcanoes in the Cascade Range in the north-west USA – erupted spectacularly. A painting of the event is shown in the colour plate section and not much imagination is needed to see the eruption cloud as a giant, long-winged bald eagle, a creature with which local people were familiar. The head and beak of the eagle appear orange-red, the cloud of steam rising from the volcano's crater (perhaps fortuitously backlit by the moon); the body and feathers of the giant bird are darker yet clearly discernible. The eagle's beak appears fastened to the mountain, as though tearing at living flesh, rivulets of blood running from a creature tethered there, its roars of pain heard by all. For non-literate observers accustomed to animistic interpretations of such phenomena, an aggressive eagle analogy was obvious.

In other parts of the world where active volcanoes are found, there are also myths about giant eagles attacking people confined on mountaintops. A well-known example is that of Prometheus, who stole fire from the gods in Greek mythology. Prometheus introduced fire to the human race, something for which he was punished by Zeus, who had him chained to a mountain in the Caucasus where a large eagle pecked daily at his liver (his immortality caused this organ to regenerate each night) something that was accompanied by loud noises and earth tremors (the screams and thrashing of the hapless Prometheus). It is understandable that the story of Prometheus has been interpreted as that of an eruption. There are similar mythical interpretations of eruptions from active volcanoes in many other parts of the world.[26] In some instances these myths have even pointed scientists towards unsuspected activity within human memory of ancient volcanoes, previously assumed to be long dead.[27]

Another example from the Pacific makes the link between memorable geological events and human mythmaking even more explicit. Compared to the continents, the ocean-floor crust is comparatively young. You can see this when you look at the global distribution of the indicators of geological youthfulness – like repeated earthquakes and volcanic eruptions – which are clustered in and around ocean basins, especially the Pacific. All oceanic islands began life as ocean-floor volcanoes, labouring to survive and grow despite the pressure of often several kilometres of overlying ocean water. Many failed, of course, yet others succeeded not only in building themselves up to the ocean surface, but also in growing above it, thereby forming some of the spectacularly active volcanic islands found today in the world's oceans.[28] In some places in the Pacific, the summits of active volcanoes lurk just a few tens of metres beneath the water's surface. When these volcanoes periodically erupt, the forced mixing of cold ocean water and liquid rock (heated to perhaps 1,000°C/1,800°F) makes for memorable (phreatomagmatic) explosive eruptions.

Sometimes the clouds of steam and ash from these volcanoes produce intriguing shapes (see colour plate section). Occasionally such eruptions produce islands that may survive anything from days to years before succumbing to the unceasing wave attack. This combination of circumstances is thought to underlie the stories that exist in most Pacific Island cultures about the demi-god Maui, who sailed from place to place with a giant fishhook that he used to pull fish-shaped islands up from the sea floor to the surface. Many descriptions of this process emphasise the 'island-as-fish' detail (see colour plate section). The thrashing of the fish as it was being hauled from the water agitated and discoloured the ocean surface, just as happens during such shallow-water volcanic eruptions. Research into the geographical distribution of the Maui legends in the Pacific has allowed insights into such eruptions in the time before written records began.[29]

The purpose of relating these stories is to illustrate how the evolution of human imagination may have been stimulated by witnessing dramatic natural events (like volcanic eruptions, giant waves and abrupt land displacement), which were felt to demand rationalisation because of their implausible nature and their apparent unprecedentedness. It is the same for us today. Confronted by phenomena with which we may be unfamiliar or for which we feel official explanations are inadequate, there is scope for us to use our imagination. The idea of 'aliens among us' may be considered an extreme example of such phenomena, but it sits on the same imaginative spectrum as some of our explanations for more pedestrian occurrences, such as unexpected traffic hold-ups or commodity shortages. The point here is that every culture on Earth, past and present, uses imagination to supply explanation where none is readily forthcoming from trusted sources.

Literate people – those who can read and write – are often dismissive, even it should be said sometimes contemptuous, of people and societies that are non-literate. In this sense, literacy is tyrannical, for it encourages us to undervalue our pasts – the knowledge amassed by those countless ancestors of ours who could neither read nor write. Literacy spawns arrogance. We might ask ourselves: if there is just so much information 'out there' on the internet and in libraries, and we have no hope of absorbing it all, what value can there be in investing time in trying to understand knowledge that has never been written down? However, what we typically forget in all this is that most of the world of information in which we find ourselves wallowing today was, until comparatively recently in human history, known only through speech. It was information encoded and processed and communicated orally.[30] And, of the tiny proportion that has been written down, most has commonly been dismissed as unimportant – primitive, romantic, at best symbolic or metaphorical, certainly not real history. Many scientists are now starting to feel differently, especially about those types of myth, named euhemeristic, that appear likely to be recollections by preliterate peoples of memorable events, those that future generations should know about.

The place appears dry and barren. We might consider it endlessly so. But to the people squatting in the shade of the trees adjoining the small pool of water, the landscape is neither harsh nor welcoming. Rather it is adequate, sufficiently replete with opportunities for subsistence. For the people and their ancestors have survived in such places for hundreds if not thousands of years, but the knowledge of how to do so was not acquired afresh by each new generation. It built on knowledge passed on from the old to the young, and enabled the people to face every new day more confident than they might otherwise have been that they would be able to sustain themselves and their kin from the land to which they were bound. The stories were related in the evenings, as the people bathed in the glow from the embers of the cooking fires. The first stories were always about their history, revolving around whence the ancestors had come, and who among that number was particularly esteemed and why. The stories of history ran into those of geography. They were stories about why this place is dry and whether it had always been thus; about the adjoining lands where contrasting conditions prevail; about the animals that inhabit the land

and how they might be captured and their flesh consumed; about where people might find water to drink; about places to avoid.

It could be argued that the purpose of oral communication between generations in such contexts was principally pragmatic – passing on the wisdom of the ancestors to younger people so that they might survive to one day pass on this knowledge to their children. Knowledge was key to survival; survival is the most basic of instincts for living things. With each new generation, new knowledge would be added, resulting – after a thousand years or more – in a formidable body of traditional wisdom. Research suggests that it was grandparents who became key in passing on this wisdom to their grandchildren, removing the obligation from parents who generally had other, more time-demanding roles in such groups.[31]

To engage young people and to encourage them to value and remember this information, it was often embellished – dressed up in arresting clothing. Exaggeration and mythmaking were common, especially in detailing the exploits of distant ancestors – people were giants who strode across the land, shaping it, taming it and making it fit for others to occupy. The reinforcement of group identity was also important for ensuring the cohesiveness of the group. This is something that was often achieved through traditional practices – the group's unique ways of doing things. These ranged from spiritual beliefs and worship protocols to more mundane things like food preparation and even forms of greeting. Finally, the medium of communication of knowledge transfer from old to young was commonly designed to maximise the listeners' attentiveness, ensuring that they would remember the message the next day ... and long thereafter.

From numerous observations of traditional storytelling of this kind in many cultures across the world, we know much about how messaging was enhanced through performance. Traditional storytellers rarely sit still. Often they stand, mimicking the actions of the characters whose exploits they are relating. Sometimes they dance or clown, making their audience laugh and encouraging its attention to the narrative. Sometimes they masquerade, wearing masks or dressing in ways that caricature those they are describing. Sometimes the power of the narrative alone is enough to captivate the listeners, in the same way as many of us today can be captivated by a gripping film or novel. For as most successful storytellers know, whether they be novelists or screenwriters, as a species we seem to have evolved a weakness for narrative. Yet what we often today label escapism may well have developed in our species from simple pragmatism. Our modern predilection for narrative may derive from our attention to survival stories.

This is not everyone's view, of course. Many people believe that every person has a 'creative streak', implanted eons ago within our genetic make-up, which explains our cumulative achievements in art and music, for example. It may also explain why we are curious and innovative, why we have 'advanced' so much compared with other species less fortuitously endowed. I have a less conjectural view. It seems to me that tens of thousands of years ago, our ancestors were never arbitrarily creative. Everything they did was purposeful because they had to spend so long acquiring the food they and their dependents needed to survive. Thus early art – cave paintings, for instance – was designed to supplement memory in order to render more effective the understanding of history and geography by each successive generation.[32] As that purpose inevitably declined in urbanising societies, so such forms of art took on a life of their own, culminating in artworks that even lack obvious meaning, at least to the untutored eye. In a parallel way, storytelling evolved in non-literate cultures to pass on practical information, but in order to convince every new generation of its importance (and ensure that it passed it on to its children), it was made more memorable through exaggeration, identity reinforcement and exciting methods of communication. The pragmatic roots of such storytelling have been lost in most of today's literate cultures, so it is only the other things that remain: narrative, drama and performance for their own sake.

Ten thousand years ago, all those cultures on Earth that had been established for some millennia had knowledge bases that relied almost entirely on oral communication. It is likely that some of this was formalised – a body of lore that all adolescents needed to know before they could be regarded as adult – and some less so. There are likely to have been many cultural practices governing the teaching of tribal lore, usually gender specific and ranging from the simple requirement that initiates repeat key texts from memory to ones involving practical tests of understanding. The latter may have included requiring initiates to spend extended periods of time alone away from their tribe, surviving on wild foods and communing with the spirits of the ancestors, thereby validating their claim to adulthood.

The San peoples of southern Africa comprise a cultural group likely to have existed for 50,000 years or more.[33] Responsible for much of the region's rock art, the San also have formidable oral bodies of traditional lore and practise various forms of initiation, which for boys represent their future roles as hunters and food providers, and for girls focus on childbearing. Storytellers are lauded among the San people and routinely employ exaggeration and mythologisation to captivate their audiences; in one such tale, aggressors are transformed to 'black ants' swarming through the camps of the Early People, eating all they encounter. Yet many San stories involve imparting practical knowledge – they range from stories articulating the subsistence possibilities of various places, to more routine tales describing 'sightings of water sources, plants ripening, stands of fruit and nut trees, animals, tracks'.[34] Performance by storytellers is common and occasionally intensifies to frenzied trances. Much San rock art has been interpreted as images of performed storytelling and may indeed have been created by the performers, perhaps shamanic, who used painting to remind their people about a particular performance and its purpose.

Rock art elsewhere has inevitably been interpreted less circumspectly. Many of California's surviving petroglyphs and pictographs are seen as an astonishing 'record of prehistoric earthquake and volcanic activity',[35] intended to inform future generations about the hazards that lie within the apparently benign landscapes of the state. Yet if such stories, including those about Crater Lake described earlier, created the larger framework for Native American youngsters to understand the history of the place in which they dwelt, more practical everyday advice was also provided orally in ritual contexts. In North America, the nature of these varied depending on the way of life followed by particular tribes.

The Tlingit people of south-east Alaska have a set of traditions (*tamánwit*) that explain the nature of appropriate Tlingit behaviour, especially in relation to the sustainable use of the wild foods on which the people were once entirely dependent. The Huna Tlingit had a gull-egg collection practice that is adjudged as having required 'a sophisticated appreciation of … gull nesting biology and behaviour'.[36] Through oral instruction, the Tlingit were taught to remove gull eggs from nests containing just one or two eggs, no more. The science that validates this practice is that a female gull aims to lay three eggs each season, so if one is taken when there are just two laid, she will lay another two to reach that number – a total of four. If two are stolen from a nest with two, she will lay another three – a total of five. But if a nest is found with three eggs in it, the chances are that the mother gull has finished laying and could not lay more eggs even if she so desired. This collection strategy optimises the chances of each female laying three eggs while supplying extra to the Tlingit collectors, something that is designed 'to conserve local gull populations while affording a substantial and predictable harvest'.[37] We can infer that similar instances of sustainable harvesting of wild foods underpinned Native American use of other resources in ancient times. For example, it has been shown that at least 700 years of fur-seal hunting on the Washington coast (north-west USA) had no impact on seal populations, plausibly because of (now-lost) oral traditions governing people's sustainable harvesting.[38]

These examples are intended to emphasise the point that non-literate cultures utilised oral traditions for a number of practical purposes, ranging from the sustainable use of particular resources to wider instruction in lore for survival. Many oral traditions also focus on history, begging the question as to why societies primarily concerned with surviving needed to know about this, apparently irrelevant, topic. Perhaps it was for the same reasons as we today learn history. It counters the forces of globalisation and cultural homogenisation by giving us a unique identity. It celebrates the unique journeys that we and our ancestors undertook to reach the place we are in now, even though that place is shared by other people whose ancestors reached it along different routes. Ultimately, history can transport each of us to a purer, quieter place where, as Ralph Waldo Emerson wrote, there are voices that 'speaketh clear' to us.[39]

The oldest known rocks on the planet we inhabit formed almost 4,500 million years ago; life appeared at least 3,500 million years ago; human ancestors (hominids) almost five million years ago and our species, *Homo sapiens*, around 200,000 years ago. We began wearing long trousers about 3,000 years ago.[40] History is a subjective business – you must choose your range, for nobody can hope to synthesise it in its entirety.

You also need to be clear about your information sources. Most historians define 'history' as observed history, validated by having been written down shortly after the events recounted took place. Then there is a time dominated by oral communication between generations, a process that some will tell you is fraught with dangers around obscuring and omission of accurate detail, which is why it is sometimes tendentiously labelled 'prehistory'. There is then an even earlier history, extending backwards through a time when humans existed but from which no human memories of things that happened have been preserved, a time that can be interrogated only by science. Tentatively we might call this 'inferred human history', because science cannot reconstruct human thoughts, so what we imagine might have happened and why may actually be completely wrong. Then with more éclat, we drift into 'geological history' – most of our planet's history – where hard impersonal data acquired through often centuries of deductive enquiry have allowed us to build convincing impersonal models of how things once were and how they became what they are today. Much of this type of history is inescapably vague because the time periods we are dealing with are so long. How do we process what might have happened in a million years? We cannot do so readily and the temptation to conflate time – to treat a million years as a thousand or even a hundred – is something that often proves difficult to resist.

The main focus of this book is a slice of the grey area within prehistory and inferred human history, a time from the coldest part of the last great ice age, about 20,000 years ago, to 1,000 years or so ago, when at least a few societies in almost every part of the world had acquired some degree of literacy. This grey area was dominated by history communicated orally and involved human societies in every part of the inhabited world. So how accurate is oral history compared to the other types?

Most people to have written on this subject, historians and anthropologists in particular, have been quite rude about oral histories, focusing on their manifest limitations rather than on their potential as ways of helping unravel the history of the distant past. To be generous, the reasons for such rudeness are easy to understand. They include the tyranny of literacy (mentioned earlier) – the idea that the written word, by virtue of rendering information essentially static over long time periods, appears naturally superior to the dynamic stories told by non-literate people. Another reason is the belief that human memory simply does not have the capacity to retain sizeable bodies of knowledge without aides mémoires like books and computer monitors, or even notepads. What is this belief based on? It is based on the way we are today, not the way our ancestors were a millennium or more ago. On the face of it, it still seems reasonable – after all, who of us can store more than a few telephone numbers in our heads? But if we had no choice, if we needed the information in order to survive, perhaps we might acquire the ability to carry even a small telephone directory around in our minds.

The point here is that we cannot readily measure the cognitive abilities of non-literate people in a non-literate world by our own – the abilities of literate people inhabiting a literate world. Naturally we would be inclined to undervalue those of our non-literate forebears. We see such bias in hundreds of instances, which have led to an orthodoxy that memories communicated only orally mostly endure only a few hundred years. But what if we could prove that they could last far, far longer? That would stir things up. That might force us to re-evaluate our assumptions about our non-literate ancestors.

The main purpose of this book is to demonstrate that there are oral histories about rising sea levels from Aboriginal Australia, and plausibly cultures in many other parts of the world, which have endured several thousand years – more than 7,000 years in the Australian case – and that they should therefore compel us to shrug off our time-hardened scepticism about how long such traditions might endure, and look with a less critical eye at some of the others that remain extant. The implications of doing so are immense, for not only does it restore some credibility to the reputation of methods of oral communication in non-literate cultures, but it also hints at hitherto unsuspected depths of the memories of our own species.

Chapter 2 explains the societal context of Australian Aboriginal storytelling – how a people's beliefs and history shaped their reality in ways that contrasted sharply (and painfully) with those of the British when they began to settle Australia in 1788. Chapter 3 describes the 21 known groups of stories about coastal drowning along the Australian fringe, and the details of the process they describe and its often memorable life-changing consequences.

The way in which the ocean surface has changed – by a rise of more than 120m (400ft) – since the coldest time of the last ice age is described in the first part of Chapter 4. The use of this knowledge to determine minimum ages for Aboriginal Australian drowning stories is explained towards the end of the chapter.

Shifting the focus away from Australia, Chapter 5 examines the nature of comparable drowning stories from other parts of the world, particularly in north-west Europe and along the fringes of the Indian subcontinent. Ancient stories other than those describing coastal submergence are outlined from different parts of the world in Chapter 6 – stories describing meteorite falls (like that in the Italian Sirente discussed earlier), volcanic eruptions, abrupt land movements and even the nature of animals that became extinct long ago.

The final chapter asks whether we have underestimated ourselves by denying for so long that oral traditions may preserve knowledge across a sweep of human history far grander than we suspected, and suggests where the new frontiers of knowledge in this field might lie.

CHAPTER TWO

Words that Matter in a Harsh Land

Not everyone everywhere knows as much about Australia as do Australians, although a dispassionate observer of world geography might rightly wonder why. For Australia is massive – pretty much the same size as Europe or the conterminous United States (Figure 2.1).

About one-third of the continent of Australia is real desert. Most of the rest is also quite dry, making Australia, after Antarctica, the world's driest continent. For humans bent on settling Australia, its arid heart in particular has long proved a challenge, yet the first arrivals not only traversed it repeatedly but also occupied it, establishing desert cultures that flourished for tens of thousands of years. We know little of these first people's journeys of exploration, their successes and failures,[1] but there is a far more complete record of European exploration of Australia's forbidding interior that allows a taste of such encounters.

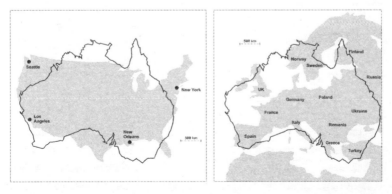

Figure 2.1 *The size of mainland Australia compared to the sizes of the conterminous United States (left) and Europe (right).*

Consider one of the first such explorers, Charles Sturt, who – buoyed by rumours (unfounded) of an inland sea – travelled across south-east Australia in 1828–1830, reporting on the Murray and Darling river systems that dominate its drainage. Sturt barely had a good word to say about anything, it seems, finding little water (much of what he found was too salty to quench the thirst),[2] and the plains 'dreary', barren and seemingly limitless. Sturt's apologetic judgement of Australia's interior was that 'there is no life upon its surface, if I may so express myself; but the stillness of death reigns in its brushes, and over its plains'.[3] Yet his belief in an inland sea lingered and he led another expedition into Australia's dry heart in 1844–1845, reaching its geographical centre in the Simpson Desert. But there was no trace of the inland sea and, battling scurvy, he headed home.

In part, Australia owes its conspicuous dryness to its topography – it is the world's lowest continent, with barely any land more than 1,000m (3,280ft) above sea level – which in turn is explainable by its geological history (see colour plate section). Currently the best model to explain the evolution of the Earth's surface is one – called plate tectonics – that pictures the entire Earth's crust as broken into a jigsaw of rigid interlocking pieces named plates, each of which is moving relative to the others. Along plate boundaries, adjoining plates may converge (often forming mountain ranges, causing volcanic and earthquake activity) or diverge (generally associated with rifting and volcanism). Sometimes two plates simply slide past one another, sticking, then periodically slipping and causing earthquakes.

Uniquely among the world's continents, Australia lies wholly within the middle of a crustal plate, far from the main sites of geological activity where plates are created or destroyed in paroxysms of seismic and volcanic fury. And, because Australia has been so located for around 34 million years, there has been ample time for denudation to work on removing its surface irregularities, unchallenged in almost every part by the mountain-building processes that have created young topography on other continents. Yet it would be wrong to imagine that distance from plate boundaries renders the Australian continent wholly immune to geological disturbances. The continent is criss-crossed by faults along which stress from movements at distant plate boundaries may occasionally shimmy. The real wake-up call in this regard was the Newcastle Earthquake on 28 December 1989, a magnitude 5.6 event in a place long regarded as seismically inactive that prompted a new and improved awareness of earthquake risk across the continent.

While most geological activity on Earth is concentrated along plate boundaries, there are places in the middle of plates where magma reaches the Earth's surface from within. Named hotspots, these places are commonly sites of active volcanoes and – most importantly – are usually stationary over long periods of time. This means that when a plate moves across a fixed hotspot, you get a succession of volcanoes being built: the youngest one sits over the hotspot, while the progressively older extinct ones trace the line of ancient plate movement. The best-studied example of such a hotspot chain is the Hawaiian Islands, in which two active volcanoes (Mauna Loa on Hawai'i Island and underwater Lo'ihi) lie above the hotspot; then, stretching away to the north-west for some 5,800km (3,600 miles), is a line of ancient volcanoes, each progressively older, which mark the passage of the Pacific Plate over this hotspot for an incredible 65 million years.

There are also hotspot chains in eastern Australia, formed when the plate of which the Australian continent is part moved over a series of fixed hotspots. The longest such chain is the Cosgrove Track, displaying a record of volcanic activity from nine to 33 million years ago (see Figure 6.1). But if the Cosgrove volcanoes are long dead, the same cannot be said for a cluster in the south of the continent, the youngest of which is Mt Gambier. Once thought to be of hotspot origin, recent research has proposed that this volcanic activity may instead have been caused by the reactivation of a long-dormant fault system along which magma has fortuitously been able to reach the surface. Yet such activity has barely affected the form of Australia as a whole. Today, being distant from plate boundaries and relatively unaffected by what happens there, the main cause of landscape change in Australia is erosion. This slowly lowers the ground surface and irons out its irregularities, creating massive quantities of detritus that fill those that remain, and thereby creates the vast, apparently featureless plains that dominate the interior.

Within the last 34 million years in Australia, the work of denudation – or land-surface lowering – has been assisted by the continent's inexorable move into increasingly warmer climate zones. At the start of this period, recently separated from the continent of Antarctica (part of the ancient Gondwana continent), Australia was located in the very cold and dry high latitudes of the southern hemisphere. Since then it has steadily moved north into lower latitudes.[4] While punctuated by wetter episodes, which were comparatively short-lived, the aridity that currently dominates Australia's climate started to affect the continent about 10 million years ago, when it reached the horse latitudes: those zones of the Earth where the high-pressure atmospheric conditions result – for most of the time – in warm weather, weak winds, few clouds and above all little rainfall.

You might therefore expect that the surface of the Australian continent has been little altered from the time it first formed – but of course this is not the case. Time is the great leveller, for the Earth-forming processes of weathering and erosion, however slow they may be, are relentless. Across almost countless eons, driven by changing temperatures, wind and rain, and running water, the land surface of Australia has been worn down. A lack of processes such as volcanic eruptions and land uplift, which elsewhere in the world rejuvenate ancient land surfaces and cause lost nutrients to be replaced, has led to a situation in Australia where soils have been weathered and leached repeatedly.

Being dry, low and somewhat slothful as moving continents go does not sound like a great recommendation for the study of Australian geology, but the reality is quite different. Australia is composed mostly of continental (not oceanic) crust, and is home to some of the earliest-known rocks on Earth, formed almost 4.5 billion years ago, outcropping in the Jack Hills of Western Australia.[5] In the same region, the remains of some of the very first life forms to appear on our planet are found. Known as stromatolites, they lived in colonies and built structures using carbonate extracted from seawater. There are living stromatolites in Shark Bay, yet a few hundred kilometres away they are found as fossils in rocks that formed 3.5 billion years ago, intriguingly close to the time Planet Earth came into being.[6]

Beneath the blanket of weathered sediment that covers most of the Australian landscape lies a complex pattern of bedrock testifying to the billions of years the continent has existed.[7] Some of this bedrock represents the cores of ancient continental masses (cratons), while other bedrock consists of remains of the sediment-choked basins that accumulated around them. About two-thirds of Australia's basement formed more than 500 million years ago.

Big land masses provide more opportunities than smaller ones for terrestrial life. Newly arrived organisms can spread out in many directions, possibly multiplying uninhibitedly at first, utilising the food sources they encounter. Food species might renew themselves indefinitely while predator population densities often remain low, below an area's carrying capacity. Such a scenario is thought to explain why the first Australians – ancestors of today's Aboriginal peoples – were able to successfully follow a largely nomadic way of life for more than 50,000 years; no other was needed to survive. Yet the climate of Australia also localised the possibilities for life that might otherwise be expected from its great size. Rainfall is concentrated along the continent's fringes, leaving a vast arid interior that poses many challenges for living things – including humans. Winds that blow onto Australian shores, often laden with moisture from their cross-ocean travels, drop most of it within a few hundred kilometres of the coast, then continue blowing inland. They are dry and pick up dust as they howl across the great inland plains to the furthest inland places where, finally, the wind losing power, the dust drops to the ground, driven into dunes in the sandy deserts.[8] Some of the deserts are stony – this is where renewed winds have removed the finer particles, leaving behind those it cannot readily shift.[9]

Compared with continents of similar size, there are not many large rivers in Australia, especially in its western half. In the centre, dry riverbeds connect series of dry lake basins that only rarely become filled with water. On such occasions, usually when a La Niña event is in progress, the desert turns green, its livelihood possibilities burgeon and its living inhabitants rejoice. When some of these lakes are brimful with water, their surfaces may become very flat and they may become almost indistinguishable from the sky when the sun is high: a phenomenon that may have given rise to legends about a vast inland sea. On his last expedition, so convinced was Charles Sturt that he would find the inland sea, he took a boat along so that he would be able to paddle across it once he found it. On 13 May 1845, an elderly Aboriginal person stayed in Sturt's camp and was 'greatly attracted by the Boat … the use of which he evidently understands'. Mesmerised, Sturt reported that this man 'pointed directly to the northwest as the point in which there was water, making motions as if swimming and explaining the roll of waves, and that the water was deep'.[10] Sturt's informant is likely to have been describing the Indian Ocean, although maybe memories of times when the dry lakes periodically filled clouded his thoughts. There has never been a sea in the centre of Australia.

The earliest known traces of human settlement in Australia date from some 40,000–60,000 years ago and are not found along the modern coast. They are found inland, typically beneath rock overhangs or in shallow caves where people once sheltered, commonly hundreds of kilometres up river valleys from the modern coastline. So of course these places cannot be the first that people settled in Australia. They must have set foot on the land hundreds, perhaps even thousands, of years earlier than the time they reached these inland shelters, but those first footprints are forever lost to us today.

This is not simply because no earlier indications of a human presence in Australia have been found along its coasts, but also because at the time people arrived here first, the sea surface was much lower than it is today. The shore on which the first people landed is now under tens of metres of ocean water, probably buried by sediments washed off the land, and perhaps even grown over with reef.

So can we know where and when the first Australians actually became Australian, or when the first human footprint appeared on Australian shores? We can use various methods to get approximate answers to these questions, including the reconstruction of the easiest series of sea crossings from Asia, or genetic tracing or even historical linguistics, which tell us where in Australia the majority of extant Aboriginal languages originated. All this is discussed later, but it is sensible to begin our quest for the first Australians by looking at the places where they first manifested themselves.

Determining ages for the earliest human settlement sites in Australia suffered for a long time from being beyond the reach of radiocarbon dating – approximately 40,000 years – but within the past 25 years newly developed techniques have been applied to diagnostic material from ancient sites that allow their ages to be accurately determined. Several such early sites are discussed below; their locations are shown in Figure 2.2.

In northernmost Australia are found a number of ancient sites, notably the two rock-shelter sites of Malakunanja,[11] and Nauwalabila (in the somewhat alarmingly named Deaf Adder Gorge). Dates for the deposition of sand containing the oldest human-made stone tools (artefacts) at Nauwalabila range from 53,000 to 60,000 years ago,[12] while those at Malakunanja are sandwiched between sediment layers dated to 45,000 and 61,000 years ago respectively.[13]

The earliest Australians used tools they made from stone, and perhaps shell, which they would have discarded when they ceased to be functional (or cherished) – like we often throw away things we no longer need or want today. Stone tools often endure the ravages of millennia, and should we

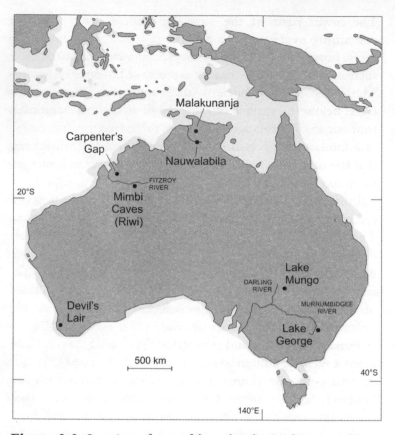

Figure 2.2 *Locations of some of the earliest known human settlement sites in Australia and New Guinea first occupied at least 40,000 years ago. The lighter-grey shaded area was dry land at the time people first arrived in Australasia from island South-east Asia about 65,000 years ago, when the sea level was about 70m (230ft) lower than it is today.*

find them today, we can often work out how long ago they were used – and thus approximate the time when their users occupied a particular place. So much of the calibration of ancient tool-using human history in Australia and elsewhere has come from working out the chronologies of tool evolution.[14]

The Devil's Lair site in south-west Australia is within a large limestone cave, 5km (3 miles) from the modern coast.

The lower parts of the cave are packed with sediments containing evidence for a human presence, including stone tools, bones and – critically for precise age determination in such ancient contexts – charcoal preserved in ancient hearths. The oldest (deepest buried) of these is found almost 3.8m (12ft) below the surface of the cave floor and has been dated to about 45,470 years ago. However, there are other indications of a human presence in layers below this hearth, suggesting that the first human occupation of this cave was at least a few thousand years earlier.[15]

Limestone caves are also sites of early occupation within the catchment of the Fitzroy River in north-west Australia. The Mimbi Caves include a cave named Riwi in which at a depth of just 65–70cm (25–28in) were found two hearths dated to more than 40,000 years ago. At the Carpenter's Gap rock shelter, 200km (125 miles) away, the sediment fill is full of well-preserved charcoal, showing that human occupation occurred here about the same time.

Finally, on mainland Australia there is the remarkable discovery of a human skeleton on 26 February 1974 in a sandbank on the shore of Lake Mungo (part of the dry Willandra Lakes system). The body had been laid in a shallow grave, then sprinkled with powdery red ochre before being buried. The first studies to date the bones used various methods and suggested that this man – Mungo Man – died sometime between 61,000 and 62,000 years ago.[16] More recent research shows conclusively that this was an overestimate. The earliest unequivocal signs of a human presence in the Lake Mungo area are from sands in which stone flakes – which could only have been created by people deliberately banging one piece of rock against another – are found; they bracket human arrival in the vicinity at between 45,700 and 50,100 years ago.[17]

The first Australians used fire, burning wood to form ash and charcoal. Some fires were wide ranging, intended to clear woodland of its undergrowth to flush out animals, expose vegetation and stimulate regrowth to which prey animals would then be attracted. Other fires were for cooking, the ubiquitous hearths or fire pits, the remains of which are often found buried within the sediment fills of caves and rock shelters where some of the first Australians once lived. As is the case with the remains of any carbon-rich life form, the age of charcoal can be determined precisely using radiocarbon dating, giving us insights into exactly when a particular tree was alive: often a reliable proxy for the time when the humans who burnt it occupied a particular place.[18]

As can be deduced from the occupation ages of inland sites discussed above, it seems likeliest that people occupied most of Australia 40,000–60,000 years ago, but there have been suggestions that their arrival was far earlier. The most common type of supporting evidence is that from concentrations of charcoal within long sediment cores from lake beds or the shallow ocean floor, which point to an unusually high incidence of bush fires at a particular point in time, suggesting (yet not proving) the arrival in a landscape of people who burnt the vegetation in order to access food.[19]

Such an idea was mooted in 1997 for the Queensland coast following analysis of an offshore core,[20] but was more famously proposed earlier when sediments from the floor of Lake George, just outside Canberra, Australia's capital city, were analysed. Lake George is usually mostly dry these days, but its form shows clearly that it was once a lake. A core 72m (236ft) long drilled into its sediment fill allowed scientists insights into its environmental history going back at least 4.2 million years. For our purposes, the critical layer (Zone F) is packed with charcoal – in stark contrast to most of the rest of the core – and was dated as having formed on the lake floor about 120,000 years ago.[21] Huge excitement greeted this discovery as commentators interpreted it as implying the first human occupation of Australia to have been far, far earlier than anyone else had hitherto suggested. Yet this age is an outlier among early proposed ages for human occupation of Australia, and for this reason is today widely regarded as wrong. Had it been correct, it would be expected that confirmatory ages from other sources would have since been forthcoming, which is not the case.[22]

A similar story of early human occupation comes from what is today Papua New Guinea, which was – because of a lower sea level – contiguous with the Australian mainland for most of the time people have lived there. A key site is the Huon Peninsula in the east of the main island. Due to its long-term uplift, this peninsula appears like a giant staircase, each step representing a coral reef that formed at a particular time. Stone axes with distinctive 'waists' for easy handling are found in sediments associated with the reef that formed at least 40,000 years ago, showing that people using this comparatively sophisticated technology were present in the area at this time.[23]

Thus while there is solid evidence suggesting that people were living in Australia and Papua New Guinea 40,000–60,000 years ago, most of the sites where their presence has been detected lie well inland of the modern coast. This seems a bit odd if you forget that, at this time, the sea level was much lower than it is today, so that the Australian coastline would have been much further out to sea. Australia would have been bigger, connected to the main island of New Guinea and closer than it is today to the islands of South-east Asia, from which we know the first Australians came. Perhaps these first settlers started their journeys on the coasts of Borneo or Java, 'hopping' by boat from island to island across the straits to reach the Australasian mainland. For this reason we might expect to find the oldest sites in Australasia somewhere in western New Guinea or northernmost Australia.

This inference seems borne out by the early age – perhaps as much as 65,000 years ago – for the occupations at Malakunanja and Nauwalabila, but even these sites would have been significantly inland at the time of first settlement, so that they were likely to have been occupied long after people first set foot in Australia. Just how long the lag was is of great interest – and worth a bit of conjecture. Given that the first arrivals were probably accustomed to feeding themselves from coastal ecosystems, it is plausible to suppose that settlement of the interior did not become a priority for several generations, perhaps only after coastal population densities increased, thereby increasing pressure on proximal food resources. It seems more likely that we are looking at a few thousand rather than a few hundred years, so if we accept that the earliest dated occupations are close to 60,000 years ago, then a time for initial human arrival of 65,000 years ago appears plausible.[24]

Many of those who have studied Indigenous Australian cultures argue that for most of the time they have existed in Australia, they remained isolated from the rest of the world's people. In almost every part of the continent, Aboriginal cultural practices and languages show clear signs of having evolved without the outside interference – the cross-cultural syncretism – that can often be so readily unpicked in cultures elsewhere. The reason for the extraordinary isolation of Australian Aboriginal cultures for so many tens of millennia is likely to lie in geographic isolation – and the fact that ocean gaps proved more formidable barriers to potential migrants than they evidently had for the first arrivals. Such inferences have been confirmed by studies of DNA in skeletal remains.[25]

In 2012, a discovery was made that shook the foundations of the long-held belief in the cultural isolation of Australia. When the first written accounts of Australia appeared in the late eighteenth and nineteenth centuries, many made mention of a distinctive wild, pointy-eared dog – the dingo – that was used by Aboriginal people as a hunting companion (although latterly it has become the scourge of sheep herds). At the time of European colonisation of Australia in 1788, the dingo was everywhere on mainland Australia except the offshore island of Tasmania. This observation proved critical to understanding when the dingo arrived in Australia. It could not have been, as was once assumed, as the ocean-going partner of the first human arrivals, for then it would surely have joined them in Tasmania, crossing the Bass Strait land bridge when the sea level was lower than it is today. The last time the ocean rose across the Bass Strait, cutting off Tasmania and its human inhabitants from the rest of Australia, was about 14,000 years ago.[26] We can therefore infer that, give or take a few millennia, the dingo was not present on mainland Australia at this time.

This intriguing scenario, which implies that the dingo must have been a significantly later arrival than the first Australians, is borne out by research into dingo DNA. The diversity of this within modern Australia is consistent with the dingo being introduced sometime between 4,600 and 18,300 years ago.[27] Put this conclusion alongside another that argues for an episode (well before European contact) of 'substantial gene flow' between Australian Aboriginal people and those of the Indian subcontinent, and it is tempting to imagine a small fleet of people and their companion animals (including dingos) sailing south-east from Tamil Nadu, headed perhaps for Sri Lanka or Indonesia, being swept off course and fetching up on Australian shores. While the dingos multiplied in this new land, the human migrants became absorbed into Indigenous Australian cultures, eventually becoming indistinguishable from the original occupants except for the telltale signs in their DNA.[28]

More is known about a later period of pre-European contact: that between the Aboriginal peoples of northern Australian coasts and the Macassar people of Sulawesi Island (Indonesia), who came in fleets bent on collecting sea cucumbers (*trepang*) – a delicacy in great demand in China – beginning around 1637. Evidence of their cultural impact comes from the representation in Aboriginal (Yolngu) rock art of the distinctive *perahu* (vessels) of the Macassar but, despite setting up processing stations for boiling and drying the sea cucumbers, they left little long-term imprint on Aboriginal Australia.[29]

When we talk of isolation and contact in an Australian context, we need to remind ourselves of scale. The continent is vast, the great majority of its Indigenous peoples remained largely isolated from the rest of the world for tens of millennia, and the local contacts had generally localised cultural impacts; that is, at least, until the arrival in 1788 of the first group of foreigners bent on staying in Australia – the British.

To understand the context, it helps to step back a few hundred years.

It was 13 February in the year 1772 and two French ships – the *Fortune* and the *Gros Ventre* – moved cautiously through the morning fog in the southern Indian Ocean when, unexpectedly, land was sighted. This was a cause for celebration, for the ships' commander, Yves de Kerguelen-Trémarec, had been charged by King Louis XV of France to find the land mass, *Terra Australis*, which many European thinkers of the time believed lay in the Southern Ocean. The land found by Kerguelen looked bleak and was clearly an island, uninhabited and obviously devoid of the economic possibilities the French hoped *Terra Australis* might yield for their benefit. But Kerguelen named it after himself and – despite failing to land – confidently declared it an outlier of *Terra Australis* and hightailed it back to France to report his landmark discovery to his King.[30] Delighted no doubt that France had discovered what explorers from other European nations had hitherto failed to find, Louis sent Kerguelen back to his eponymous island to make its link to the fabled southern continent explicit. Unfortunately, no amount of imagination could colour this canvas, and Kerguelen returned to France in 1775, confessing the isolated island to be 'as barren as Iceland, and even more uninhabitable and uninhabited', and clearly no continent either, information that incensed Louis and led to Kerguelen's imprisonment.

Where had this belief in a vast southern continent, *Terra Australis*, pregnant with economic promise for expansionist eighteenth-century European powers vying with one another to build the largest global empire, come from? As far as we know, the belief originated with the Greek mathematician Eratosthenes who – more than 2,000 years ago – calculated the size of the *oikoumenē* (inhabited world) to be just one-quarter of the size of Planet Earth. Since the Earth rotates uniformly rather than wobbling, he reasoned, the supposedly heavier continents must be evenly distributed. Thus there must be three other chunks of land of similar size elsewhere on the

Earth's surface that balanced the distribution of land and sea. The southern hemisphere was effectively unknown to classical scholars of this period in Europe, so they made maps showing a vast southern continent (named Antichthones), which they regarded as inaccessible because of the apparent impossibility of traversing the equatorial regions separating it from Europe.

While many such ideas became sidelined in Europe for about a millennium following the decline of the Roman Empire, they were enthusiastically rediscovered by Renaissance thinkers and aspirant explorers between the fourteenth and sixteenth centuries. Some voyages of exploration at this time led to European discoveries of southern lands that were taken as corroboratory evidence of the existence of what became known as *Terra Australis*. These included the landing on Tierra del Fuego in southernmost South America by Ferdinand Magellan in 1520, the earliest written account of New Zealand by Abel Tasman in 1642, and of course Kerguelen's incautiously overstated description of his eponymous island in the 1770s.

The shattering of the belief in *Terra Australis*, and the attendant abandonment of the Greco-Roman belief in Antichthones that underpinned it, came when Captain James Cook anchored the *Resolution* at Spithead (England) at the end of his second circumnavigation on 29 July 1775 and made haste for the Admiralty in London. Cook reported how his ships had repeatedly traversed the Pacific Ocean, from east to west and from north to south, and had found no trace of a large continental mass answering the description of *Terra Australis*. The verdict was accepted, and a chapter in the history of European exploration and understanding of the world closed.

Of course, the situation might have been quite different had the continent of Antarctica, which is located pretty much where Eratosthenes placed his Antichthones, been discovered earlier. But the dangerous seas and inhospitable climate in the southern polar regions put off many potential explorers, and it was not until the 1820s that anyone landed on the ice shelf surrounding this frozen southern continent. The realisation that it fringed a huge, ice-covered continent came decades later, by which time all thoughts of the *oikoumenē* as a serious explanatory tool for understanding the configuration of land and ocean on our planet had been abandoned.

Even though its first recorded circumnavigation took place in 1802–1803, Australia was never seriously taken to be *Terra Australis* – it stretched too far north and not far enough south. But for Matthew Flinders in 1814, this was evidently as good as it was going to get, so he named the land Australia. By this time, a British settlement had been established for 26 years at Sydney Cove, today part of Australia's largest city. Britain had declared Australia to be *Terra Nullius* (no-one's land), available for the taking, a ruling used to justify land grabs and the displacement or worse of Indigenous Australians from places their ancestors had occupied for millennia. As is the case with any large-scale cultural contacts, those between Australian Aboriginal peoples and European settlers were varied, but dominated by ones that were confrontational in which the former generally lost out. Yet within the first 100 years or so of European arrival in Australia, there were also instances where settlers learnt to respect Aboriginal peoples' wisdom and familiarised themselves with it. In several instances this led to written compilations of contemporary Indigenous knowledge, including many of the stories described in Chapter 3.

Colonisation and globalisation both played a role in the subsequent subordination of Aboriginal cultures in Australia, and it is inevitable that much of the traditional knowledge they possessed before 1788 – which had been accumulated over tens of millennia – has been forever lost. Yet the part of it that remains is sufficient to allow the impartial observer insights into the impressive scope and longevity of ancient Aboriginal wisdom, foremost perhaps among that to which we, as a species, have authentic access today.

When we look back in time from the high point that human culture has attained, it is tempting to suppose that our early ancestors lived simple lives, most of which were spent acquiring sufficient food to feed themselves and their dependants, and that they had no time for what we today might describe as leisure pursuits. Half a century ago, this view pervaded much scientific thinking about early human history, but in recent decades there has been ample cause to re-evaluate such ideas. One example is provided by the inescapable conclusion that, as discussed in Chapter 1, the ancestors of the first Australians must have crossed a succession of ocean gaps as much as 70km (43 miles) across in order to reach Sahul (Australasia) from Sunda (South-east Asia). Such trips could not have been accidental or unplanned. To have succeeded, they must have required knowledge of watercraft construction and navigation, and even – you would think – some anticipatory planning for surviving when a foreign shore was reached. Such enterprises required cooperation between people, with particular individuals being assigned different roles. This would have entailed language and a degree of human cognition comparable to our own for people 65,000 years or more ago – something that Western science once viewed as impossible.

Another area in which scientists have traditionally underestimated the ability of our ancient ancestors concerns their artistic abilities. Until the challenges associated with reliably dating rock art were effectively overcome some 25 years ago, claims of its great antiquity were invariably treated sceptically. Yet with the advent of radiocarbon dating using accelerator mass spectrometry (AMS) techniques capable of determining the age of tiny quantities of carbon, it has been demonstrated that surviving rock art in parts of Africa and Europe was created more than 30,000 years ago.[31] Survival is of course key and it undoubtedly requires a particularly fortuitous series of events to preserve rock art this long. In Australia, the oldest dated rock art is a charcoal painting that fell face-down from the ceiling of the Nawarla Gabarnmang rock shelter in Arnhem Land and was then buried beneath cave sediments for 26,913–28,348 years. Mud dauber wasp nests built over the head-dress of a 'mulberry-coloured human figure' painted on a rock face in the Kimberley region allowed determination of a minimum age for the artwork of

16,400 years ago,[32] the earliest dated image currently known in Australia.

It is, however, certain that earlier Australians created rock art and that this was part of a range of symbolic expression that may even have come with the first arrivals. In support of this, worn ochre crayons made some 40,000–60,000 years ago have been found at both Malakunanja and Nauwalabila, although it is unclear what they were used for. Yet from here it is only a small step towards supposing that, in addition to their maritime and artistic accomplishments, the first Australians used language in a similar way – to celebrate and codify their culture and to accumulate an oral record of their history. It is generally supposed that at most only a few languages would have been spoken by the first Australians (perhaps even just one), but that as time went on and their descendants spread across Australia, the numbers of different languages increased. At the time of the first record, around the start of the nineteenth century, a group of languages known as Pama-Nyungan (PN) was spoken across about 80 per cent of Australia, while non-PN languages were spoken only in parts of its north and north-west.

Such a history of language development and diffusion laid the foundations for Australian Aboriginal cultures to encode important observations of natural phenomena in their oral traditions, and also to develop methods of passing these on from generation to generation in ways that proved both effective and sustainable. While (as we shall see shortly) many observations were recorded in ways that were readily understandable hundreds or even thousands of years later, other observations were mythologised, perhaps because they might otherwise have seemed too implausible – and therefore unimportant – to later generations. Thus was born the Dreaming (*Tjukurrpa*),[33] the rich and varied world of the mind within which Aboriginal people's culture has long been grounded, and which is believed to exist in parallel with the tangible one and renders the past 'far beyond the memory of any person, but conserved in the collective memory of the whole community'.[34]

Before the advent of literacy, the Dreaming of Aboriginal Australians served the same function as a library does for literate people today. It contained 'books' that could be read only by those who had been taught to 'read' them. Among those books were technical manuals explaining the practicalities of surviving in Australia's harsh environments, where to find water and food, and how to teach your children and grandchildren to read. Yet in the Dreaming Library were also books of geography that explained the character of the landscapes with which a particular community had interacted and the changes they periodically underwent. These changes might also be recounted in the history texts in the Dreaming Library, emphasis being placed on memorable events that tribal ancestors had witnessed: events that may have included volcanic eruptions, meteorite falls, extreme waves, prolonged droughts and – central to the stories in Chapter 3 – the drowning of once-dry lands along Australia's coastal fringe.[35]

Reading a book on the shelves of the Dreaming Library required one to listen to a person who kept the books in their minds. Such people would be those who were older, and who had once been inculcated with this knowledge by their elders. Ethnographic information suggests, quite plausibly, that there were formalised processes in Aboriginal peoples' cultures for teaching younger people how to read these books and retain the information in their minds, ready one day to pass on to their own offspring. These processes manifest themselves most commonly as adults having a cultural responsibility to pass on the 'law' to youngsters. Yet it is clear that in Indigenous Australian cultures there are also inbuilt mechanisms to ensure that the law as transmitted is accurate and complete; omissions or errors in content could have fateful consequences for subsequent generations. Take as an example that a man (A) teaches the law through a series of traditional stories to his son. It is then incumbent on A's daughter's son, who is taught the law through a different patriline, to check that his maternal uncle (and his children) has the stories right.[36] Such cross-generational checks for accuracy and completeness

ensured – as far as possible – that each successive generation
had the stories correct and was therefore as well equipped as
possible to survive.

The innate conservatism of Aboriginal peoples' cultures in
Australia is also likely to have played an important role in the
effective multi-generational transmission of tribal law. Many
Aboriginal people expressly state the imperative of telling a
story the correct way, something that not only ensures its
content is accurate, but also that the ownership of the story
(and with it the responsibility to pass it on through the patriline)
is made explicit. Particular family lines often have their own
stories. While these may be heard by others for whom they are
not principally intended, those people do not own these stories
and understand that they have no right to retell them.

Many books were read (from memory) out loud, and some
were supplemented by performance (dancing and clowning).[37]
Sometimes the stories in the books were sung, the tunes of
the songs prompting the singer's recollection of the words.[38]
Rock art was used to reinforce such messages, to be enduring
reminders (or mnemonics) of key details from which observers
with sufficient base knowledge might then reconstruct the
rest of the narrative. Other forms of Aboriginal peoples'
artwork had similar functions;[39] an extraordinary map of
waterholes in the central desert, initially thought to be a
purposeless, solely artistic design, is shown in the colour plate
section.[40] This map illustrates the purpose of much Aboriginal
art – as an essential aid to surviving in a harsh land.

For hunter-gatherers to survive in Australia, particularly
through times of more than usual climate-driven stress, they
needed to intimately understand the environment they occupied
and the full spectrum of livelihood possibilities it represented.
Probably this was a lesson Aboriginal societies learnt the hard
way – that anything less than intimacy might invite disaster,
perhaps in the form of the death of an entire community
through starvation or thirst. So the landscape and the climate
shaped culture, forcing it to be conservative rather than
innovative, and in doing so it ensured that formidable bodies of
traditional knowledge – the Dreaming Library – were taught

anew to every new generation as comprehensively and as accurately as they had been to previous ones.

The wisdom that underpinned the traditional ways of Aboriginal Australians was something that European colonists of Australia were generally slow to acknowledge.[41] Like colonising people in other situations who were certain that their ways of interacting with environments in order to feed themselves were superior to those of the colonised – and who had the weaponry to support this view – most of the new arrivals in Australia after 1788 became contemptuous of Aboriginal knowledge.[42] They were resolute in their belief, for example, that the kinds of seasonal interactions with temperate landscapes that had been developed in their native Europe were applicable to Australia. And why should this not be so?

The answer lies in the fact that 'Australia is the only continent on Earth where the overwhelming influence on climate is a non-annual climatic change' – the El Niño Southern Oscillation (ENSO).[43] This means that for the vast majority of the continent, annual (seasonal) cycles of change are dwarfed by those superimposed by ENSO. When, every three to five years or so, Australia enters an ENSO-negative phase (an El Niño event), drought affects most parts. In the distant past, successions of such droughts were sometimes severe enough to force temporary Aboriginal peoples' abandonment of Australia's driest inhabited parts.[44] Today they entail water stress on farms, especially within the broad transition zones separating the wetter coasts from the parched interior and, increasingly, water shortages for people in the country's densely populated urban areas. Conversely, when Australia is affected by an ENSO-positive phase (a La Niña event), things frequently become much wetter than usual; often the dry centre of the continent is given several prolonged soakings, sometime sufficient to bring forth plant life in places where one might otherwise never have suspected it to be lying dormant. From this long acquaintance with the vagaries of the Australian climate, its Aboriginal inhabitants evolved land-use practices and techniques for accessing water that were unfamiliar to European colonists, many of whom inevitably undervalued or ignored them. A couple of examples illustrate the point.

It was natural for many of the earliest Europeans to judge the potential of Australian landscapes for agricultural production through the lens of their experience. One key analogy for some of the earliest British observers of Australia was the presence of the grassland areas along its east coast. On 1 May 1770, James Cook and others made an excursion onshore from HMS *Endeavour*

> ... *into the Country which we found diversified with Woods, Lawns and Marshes; the woods are free from underwood of every kind and the trees are such a distance from one another that the whole Country or at least a great part of it might be Cultivated [for food crops] without being obliged to cut down a single tree.*[45]

This view was echoed by Sydney Parkinson, illustrator aboard the *Endeavour*, who was the first to draw the analogy between this landscape and 'plantations in a [English] gentleman's park'.[46] As more was written about the Australian landscape, so its remarkable yet puzzling patchwork nature was increasingly remarked upon. In the early 1840s, Henry Haygarth described his discovery of the Omeo (Omio) Plain in Victoria:

> *The gloomy forest had opened, and about two miles before, or rather beneath us ... lay a plain about seven miles in breadth. Its centre was occupied by a lagoon ... On either side of this the plain, for some distance, was as level as a bowling-green, until it was met by the forest, which shelved picturesquely down towards it, gradually decreasing in its vast masses until they ended in a single tree ... By what accident, or rather by what freak of nature, came it there? A mighty belt of forest, for the most part destitute of verdure, and forming as uninviting a region as could well be found, closed it in on every side for fifty miles; but there, isolated in the midst of a wilderness of*

desolation, lay this beautiful place, so fair, so smiling, that we could have forgotten hunger, thirst, and all the toils of the road, and been content to gaze on it while light remained.[47]

It was not long before more prescient writers came to recognise that these landscapes were human-made, created by Australia's Aboriginal peoples to provide them with a regular source of food. Writing in 1848, Sir Thomas Mitchell explained it clearly:

Fire, grass, kangaroos, and [Aboriginal] human inhabitants, seem all dependent on each other for existence in Australia; for any one of these being wanting, the others could no longer continue. Fire is necessary to burn the grass, and form those open forests [grassland-savannahs – the gentlemen's parks], in which we find the large forest-kangaroo; a person applies that fire to the grass at certain seasons, in order that a young green crop may subsequently spring up, and so attract and enable him to kill or take the kangaroo with nets. In summer, the burning of the long grass also discloses vermin, birds' nests … on which the females and children, who chiefly burn the grass, feed. But for this simple process, the Australian woods had probably contained as thick a jungle as those of New Zealand or America, instead of the open forests in which the white men now find grass for their cattle, to the exclusion of the kangaroo, which is well-known to forsake all those parts of the colony where cattle run.[48]

Over tens of millennia, Aboriginal peoples learnt how to manipulate the landscapes of Australia to optimise their food-producing capacity. Within a few decades, in several places European settlers had undone this relationship, seduced by the Arcadian appearance of the grassland savannahs. They paid the price as – it could be argued – modern Australians continue to do today. For Aboriginal burning of the savannahs every few years was not only designed to maintain food supply, but also to avoid much bigger natural 'burns'. Due to the climatic dominance of El Niño droughts, much of the Australian vegetation catches fire every seven years or so; it has evolved to do so. Yet the size of these natural fires depends on the amount of dry fuel load lying on the forest floor – what Cook called 'underwood' – for if there is a lot, the fire will burn fiercely and widely, threatening all living things. Aboriginal people knew this and understood that by deliberately burning these areas every two to three years, underwood would be prevented from accumulating sufficiently to fuel catastrophic bush fires.

Not able or willing to recognise Aboriginal ingenuity, many early commentators on Australia saw instead a people who were neither sedentary nor farmers – the assumed hallmarks of civilised society – so concluded the country to indeed be *Terra Nullius* (nobody's land). This provided a legal basis for colonists to drive Aboriginal peoples off lands from which they had subsisted for thousands of years; it was a short-sighted view,[49] which failed to acknowledge that nomadism was far better suited to Australia's environmental and climatic context than sedentism:

> *Nomadism was clearly an adaptation to tracking the erratic availability of resources as they are dictated by ENSO. Nomadism has a great cost, for possessions must be kept to a minimum. The Aboriginal tool kit was thus rather limited, consisting of a number of usually light, mostly multi-purpose implements. Investment in shelter construction is likewise constrained by such a lifestyle, for there is no point in building large and complex structures when ENSO may [abruptly] dictate that the area be deserted for an unknown period at any time.[50]*

Land from which Aboriginal peoples were displaced was commonly 'bought', then fenced – two alien concepts for Indigenous Australians – and used to graze sheep or to systematically plant crops. It was no longer deliberately and routinely burnt, which opened the door to periodic catastrophic bush fires of the kind that annually raze massive parts of Australia today.[51]

In addition to fire, water is the other key to sustaining life in Australia. The Aboriginal people of Australia's deserts not only had maps of waterholes (see colour plate section), but were far better equipped than later settlers to access water where there appeared to be none. One nineteenth-century story recollects that a European travelling through the Western Desert was close to perishing from thirst; 'at the last gasp, he came to a clay-pan which, to his despair, was quite dry and baked hard by the sun'. He gave up hope, but not so his Aboriginal companion:

> ... who, after examining the surface of the hard clay, started to dig vigorously, shouting, 'No more tumble down, plenty water here!' Struggling to the side [of his companion], he found that he had unearthed a large frog blown out with water, with which they relieved their thirst. Subsequent digging disclosed more frogs, from all of which so great a supply of water was squeezed that not only he and his [companion], but the horses also were saved from a terrible death![52]

Such Aboriginal people were adept at recognising telltale marks made by these water-bloated frogs on the sides of clay pans, in which they might bury themselves for more than a year. Another technique involved stamping on the surface of the clay pan and listening for the faint croaks of the aestivating frogs below its surface.[53] Yet despite such time-honoured ingenuity, the search for water by Aboriginal peoples during prolonged droughts was sometimes hopeless.[54]

We have seen how over the course of perhaps 65,000 years Aboriginal people evolved a fine-tuned understanding of how to live well from Australia's natural environment.[55] Given that the complex, area-specific information – the law – could not expect to be effectively learnt anew by each new generation, Aboriginal Australians formalised instruction that was quality assured, allowing the effective cross-generational transmission of the Dreaming Library. But how long might

this knowledge be expected to endure before its essence
became obscure?

Probably all of us have had pause to wonder at some point
about the apparent superstitions of our parents' generation or
earlier. For me, it was the seemingly meaningless imperatives
about never walking beneath a ladder leaning against a wall, or
fully expecting days, nay weeks, of bad luck should a black cat
cross my path. When I was growing up, I was understandably
sceptical, even contemptuous, of such beliefs, but now I wonder
whether in fact they might represent legacies of practical
advice, once passed down from one generation to the next as a
protection from harm. I cannot today comprehend what legacy
the ladders and black cats might denote – and I do not believe
my parents did – but of course this does not mean that these
stories do not represent such a legacy.

Every culture in the world has such legacy stories, many of
which – particularly those with spiritual dimensions – have
come to define particular groups. In richer societies where
global networks dominate the local ones, most such stories
are today mere anthropological curiosities. Yet in many
poorer parts of the world, where people depend more on
local than on global networks, such stories are important keys
to unlocking local knowledge. Consider the 3,000 people
who live on Savo Island (Solomon Islands, South-west Pacific
Ocean), the top third of an active andesitic volcano that has
erupted spectacularly three times since AD 1567. The people
of Savo have many stories about the precursors of eruption
that helped their ancestors evade its worst effects. These
stories include the filling with water of the usually dry
volcanic crater, an increase in geothermal activity and –
intriguingly – the shrinking of the island through wave
erosion of the loosely consolidated volcanic sediments
produced by the last eruption. The stories tell that once the
coastline has been eroded back to the foot of the hills, then
another eruption is due – a unique empirical chronometer for
eruption recurrence.

From this and many other examples, it is clear that the
utility of such stories and the information they contain is

context specific. As long as the people of Savo remain on their island, dependent on its abundant natural resources, knowledge of when their way of life is periodically threatened by eruption will be maintained. But as soon as people living on Savo become detached from the local, perhaps through the introduction of imported foods and digital communication networks (as is happening throughout the Pacific Islands), then local knowledge of that kind will inevitably be devalued and replaced by a dependency on external knowledge. This is likely to involve scientific ways of predicting and warning of imminent eruptive activity.

A good example of the enhanced vulnerability of comparatively isolated communities in poorer countries following loss of traditional knowledge for coping with disaster is the *smong* tradition of Simeulue Island (Indonesia). This tradition teaches that when the ground shakes, people should drop everything and run for the hills, because a massive wave – a tsunami – is likely to soon sweep across the coast. The *smong* saved countless lives on Simeulue during the great Indian Ocean Tsunami in December 2004, which involved waves of 10m (33ft) high, whereas the death toll along neighbouring coasts where there were no such traditions was far greater. In the aftermath of this phenomenal tsunami, there was a flurry of global interest in improving scientific early-warning systems of such events in order to reduce their human impact. Yet even if that communication were optimal, it would still take at least 15 minutes, probably much longer, from the time of the earthquake to the earliest time people in danger zones could receive the first warning. That would not have been much use on Simeulue in 2004, for the first wave washed over the coastal settlements just 10 minutes after the earthquake was felt. The *smong* and similar culturally embedded traditions are clearly superior for communities closest to the epicentres of such massive earthquakes.[56]

For Aboriginal Australians, occupying an ocean-bounded continent in effective isolation from the rest of the world for perhaps 65,000 years, the cultural context may have evolved only very slowly compared to other, less isolated parts of the world. This meant that the content of important stories would not necessarily attenuate in the way it might in more rapidly evolving cultures with more permeable borders. In other words, geographical isolation is key to cultural homogeneity, which in turn nurtures the development of effective and enduring strategies for coping with environmental stresses.

Yet if cultures have been remarkably static in Australia for much of the last 65,000 years or so, the same cannot be said of environments. It is probable that Aboriginal knowledge stories were adapted to environmental change, with the books in the Dreaming Library being periodically revised. Key to this is appreciating how slowly much of this change occurred, giving successive generations of the peoples of the land ample time, you would think, to adapt their stories to the changing conditions.

When we cast our gaze back across the last 65,000 years of our planet's history, the single most significant event to have occurred in every part of the world was the last ice age. Not only was it cooler than it is today, with the ocean surface being correspondingly lower, but in much of tropical Australia the coldest times of the last ice age were also drier. Already adapted to life on the second-driest continent, many Aboriginal groups evidently found ice-age aridity too much to cope with, for they are known to have abandoned some of the continent's most arid areas during the coldest and driest part of the last ice age, their former inhabitants clustering into refugia where there was sufficient water and food for them to survive.[57]

When the ice age ended, Australia's climate became warmer and wetter, and its inhabitants reoccupied almost every corner of this vast continent.[58] It is likely that most of the Aboriginal stories to have survived until today, discussed in the next chapter, were refashioned during this time of climatic amelioration to reflect the new environmental conditions and the concomitant possibilities they presented for human survival. It is also probable that some of the environmental processes that gave birth to these new conditions – like the rise of the ocean surface – became etched into Aboriginal storytelling, an essential element of the law that provided people with a context for understanding the range of new environmental possibilities. It is to these transformative processes that we now turn our attention, one type of which – sea-level rise – is the principal subject of the Aboriginal stories related and analysed in the next chapter.

CHAPTER THREE

Australian Aboriginal Memories of Coastal Drowning

O ne of the most common extant mythical characters in Australian Aboriginal cultures is the Rainbow Serpent, about which innumerable stories have been told and which has been represented in artworks, perhaps most enduringly in rock paintings, for at least 6,000 years. The original Rainbow Serpent came from the sea, and upon occupying the land had many progeny that made it their home.[1] Stories about the Rainbow Serpent tell of how it winds its way across the land, wrapping its coils around the hills, shimmying across or beneath the land, furrowing the plains with its sinuous body to create meandering watercourses. At rest, it is often portrayed as the landscape itself, its head and body recognised in the topography. Sometimes it is seen in the sky, whirling and diving, with lights emanating from its body bathing the lands below in a kaleidoscope of colour. The Rainbow Serpent is often portrayed as the guardian of the land, embodying the spirits of the ancestors, the essence of Aboriginal culture, zealous in the stewardship of its integrity, and sometimes visiting a terrible revenge on those who disobey the law.

Its singular form and behaviour have often led to the Rainbow Serpent of Aboriginal storytelling being characterised as a composite being, formed from different parts of real creatures. Yet other research suggests that it is almost an exact representation of the pipefish *Haliichthys taeniophorus*, or one of the sea horses that might have been among the more striking creatures, least familiar to local inhabitants, to have washed up on the shores of northern Australia when the postglacial sea level ceased rising 6,000–7,000 years ago. It is therefore perhaps no coincidence that this is the time at which Rainbow Serpent imagery appears in Australian rock art in its fullest developed form.[2] What such research implies is that oral traditions informed artistic expression that in turn contributes to their recollection.

The idea that the enduring tradition of the Rainbow Serpent in Australian Aboriginal cultures originated with observations of a memorable event (or events) is in keeping with ideas about the origins of many myths – those that are termed euhemeristic. It would be imprudent to claim that all myths are euhemeristic since that would leave no room for human invention, but in recent decades the euhemeristic nature of many myths has become increasingly apparent. As a result, the long-assumed fictional nature of many myths has been challenged. Part of this is due to the impartial scrutiny of particular myths by scientists, especially geoscientists, which led to the demonstration that non-literate cultures had preserved information about certain events and phenomena that science had overlooked, largely because the conventionally admissible evidence appeared fragmented and difficult to reconstruct. Some examples were given in Chapter 1 – those of the Klamath stories about the death throes of Mt Mazama, and of the people of the Sirente about a meteorite fall – but there are many more. They include the linking of the Delphic Oracle in Ancient Greece to surface emissions of hallucinogenic gases,[3] geological explanations for abrupt disappearances of islands in the Pacific Ocean basin,[4] and even the possibility that sightings of the Loch Ness Monster in Scotland were merely expressions of the effects of strong earthquakes agitating the lake-surface water.[5]

While Australian Aboriginal cultures differ from many elsewhere in the world in terms of their superior replication fidelity and the longevity of their traditions, the importance of preserving memories to allow future generations to understand their people's journeys is shared by every cultural group – literate or non-literate. Pre-colonisation Aboriginal cultures were non-literate, yet had rich oral and artistic traditions that provided vehicles for intergenerational transmission of stories. Given that these cultures have been massively altered by the

Europeanisation and subsequent globalisation of Australia, we owe much of our knowledge about their original content and depth to

> ...*curious, observant, and relatively unprejudiced individuals in all parts of Australia [who] wrote down descriptions of Aboriginal ceremonies, recorded versions of Aboriginal myths and tales and sometimes gave the texts and even occasionally the musical scores of songs.*[6]

However abhorrent we might today find the attitudes of many colonising peoples towards the indigenes – not something unique to Australia – we should yet be thankful that such 'curious' persons existed, prepared to make an effort to understand and record the wisdom of Aboriginal people before it became diluted or even forever lost. Some Anglo-Australian recorders of Aboriginal stories understood the responsibility they had for proper engagement and accurate recording. One such individual was James 'Jimmy' Dawson, who settled in Australia in 1840 at the age of 34 and, while running the Kangatong cattle and sheep station in Victoria, studied local Aboriginal culture. He described his approach in his 1881 book:

> *Great care has been taken in this work not to state anything on the word of a white person; and, in obtaining information from Aboriginal people, suggestive or leading questions have been avoided as much as possible. The informants, in their anxiety to please, are apt to coincide with the questioner, and thus assist him in arriving at wrong conclusions; hence it is of the utmost importance to be able to converse freely with them in their own language. This inspires them with confidence, and prompts them to state facts, and to discard ideas and beliefs obtained from the white people, which in many instances have led to misrepresentations.*[7]

Such fine words do not, of course, mean that every Aboriginal story discussed in the rest of this chapter should be uncritically regarded as an authentic Indigenous original – it would be naive to suppose so – or that the rendering of these stories captures all the nuances, indeed all the details, of the original oral tradition. Nor can it be assumed that some stories do not contain an overprint of European thought, imposed either by their Indigenous narrators (who had inevitably had some exposure to the colonisers' culture) or by non-Indigenous recorders. Yet even with such caveats, the fact that stories from similar locations contain much the same detail suggests that they are reporting much the same thing. And, of course, the fact that stories from at least 21 locations along the coast of modern Australia, measuring some 47,000km (29,200 miles) in length, can be plausibly interpreted in the same way – as memories of a time when the ocean rose across the coastline (and never receded) – strengthens their interpretation as recalling postglacial sea-level rise, an event around Australia that ended about 7,000 years ago.

The world was plunged into the last great ice age about 90,000 years ago. As a result of falling temperatures, in many places evaporated ocean water was precipitated back on to the land in solid form – as snow or ice. Huge ice sheets developed on many higher latitude continental areas, trapping the former ocean water, which resulted in the ocean surface (sea level) beginning to fall. Around 20,000 years ago during the coldest time of the ice age – the Last Glacial Maximum – the global sea level was around 120m (400ft) lower than it is today. Then global temperatures began rising, ice started to melt and much of the water that had been trapped in terrestrial ice sheets began to be returned to the oceans, causing the sea level to start rising. As discussed in more detail in Chapter 4, this process was neither continuous nor monotonic, but for all that it had a massive transformative effect on the world's coasts – and the peoples who occupied them. For example, one plausible estimate has it that 14m (46ft) of the coastline of northern Australia would have been (laterally) submerged every day during the more rapid periods of postglacial sea-level rise. This was surely something that would have caught the attention of coastal dwellers and made it well suited for recording in oral traditions.

Australia's Aboriginal stories about postglacial drowning are of two types, what those tireless students of Aboriginal anthropology – Ronald and Catherine Berndt – termed 'ordinary stories' and 'sacred mythology'. The first type are narratives, apparently little embellished, which may describe a time when the sea level was lower than it is today, and the shoreline of a particular part of Australia was consequently further seawards. Such narratives invariably describe what then happened, how the ocean rose, flooding familiar landscapes – places that had names and historical associations for local people – and transforming their environments and their livelihood possibilities. The second type are myths, often alluding to changes to coastal environments similar to those described in the narratives, but explaining these changes in terms of the actions of particular individuals – sometimes super human, more often non-human (like giants or god-like beings with magical powers). Both types of story can be interpreted in the same way. They report a time when the sea level rose across the land, flooding and then drowning it until it came to appear the way it does today.

A map showing the 21 locations from which stories or groups of stories have been obtained is shown in Figure 3.1. The descriptions start in the south at Spencer Gulf.

On 20 March 1802, four months into the first reported circumnavigation of Australia, Matthew Flinders, an English sea captain, steered his ship *Investigator* into a 'gulph' that was duly explored. Disappointed that this was evidently not the east coast of *Terra Australis*, that the water at the head of the gulf was as salty as the ocean and that despite abundant signs of habitation 'we had not the good fortune to meet with any of the people', Flinders contented himself with naming the place Spencer Gulf, after the First Lord of the British Admiralty at the time the *Investigator* was commissioned for this voyage.[8]

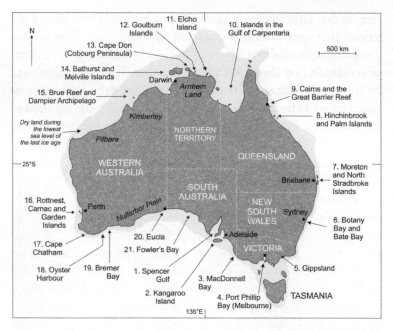

Figure 3.1 *The 21 locations along the coast of modern Australia from which Aboriginal stories about coastal drowning have been collected.*

The *Investigator* was being watched. Even if we did not know this, we would consider it likely, of course, but we do know because local people have a memory of this landmark event. The Aboriginal people occupying the western shores of Spencer Gulf – the Nawu – have a story of a beautiful white bird that 'came flying in from over the ocean, then slowly stopped and, having folded its wings [its sails], was tied up so that it could not get away'.[9]

Spencer Gulf is a triangular-shaped indentation, 300km (186 miles) in length, along the coast of South Australia, bounded on the west by the Eyre Peninsula and to the east by the Yorke Peninsula. Spencer Gulf is a graben, a fault-bounded depression that has been slowly subsiding for at least a couple of million years. Yet since the rate of subsidence – perhaps less than a tenth of a millimetre a year – is so slow, around a thousandth of the average rate of postglacial sea-level rise, it can effectively be ignored in any consideration of the stories that recall the drowning of Spencer Gulf. The Gulf is comparatively shallow, its floor buried beneath tens of metres of sediment laid down by the rivers that flow into it, as well by the periodic marine incursions – marking the terminations of ice ages of the last few million years – that have affected it. As shown in Figure 3.2, over the 300km length of Spencer

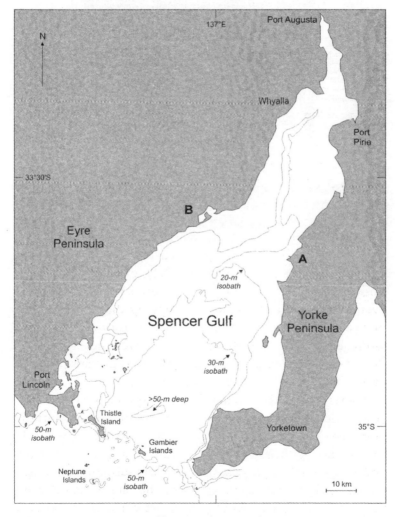

Figure 3.2 *Spencer Gulf in South Australia.*

Gulf, its floor drops only some 50m (165ft). Thereafter, when you cross its southern lip, the sea floor plunges rapidly downwards.

The shallowness of Spencer Gulf – most of it is less than 50m (165ft) below sea level – means that it was dry for tens of thousands of years during the last ice age, for most of which people were living in Australia. Some eyewitness accounts of this time have come down to us today. The stories about a time when Spencer Gulf was dry land are part of the history of the Narungga (or Narangga) people, who have lived for millennia on the Yorke Peninsula. The earliest written account dates from 1930 and tells that the Narungga 'had a story that has been handed down [orally] through the ages. It is a tale of … when there was no Spencer's [sic] Gulf, but only marshy country reaching into the interior of Australia',[10] exactly what you might expect if such a low-gradient area of land were emergent today.

The story goes that some groups of the animals that occupied dry-land Spencer Gulf had disagreements with each other.[11] The birds in particular 'felt so superior to the rest of creation that they prohibited the [other] animals and reptiles from drinking at the lagoons' that traced the axis of Spencer Gulf. Thus, the story continues, 'began a long conflict in which many were killed, and large numbers of land-dwellers died of thirst'.[12] This sounds like an analogy of a human territorial conflict, perhaps exacerbated by a prolonged drought of the kind that often affects Australia during El Niño events.

This situation caused considerable anxiety to the leaders of other groups of animals not directly involved in the conflict – particularly the emus, the kangaroos and the willie wagtails[13] – so an eminent kangaroo, after a revelatory dream, was led to a giant magical thigh bone belonging to one of his dead ancestors. He pointed this at the mouth of Spencer Gulf, causing the sea to enter it, then dragging the bone behind him to create a furrow, he walked towards the head of the Gulf. 'The sea broke through, and came tumbling and rolling along in the track cut by the kangaroo-bone. It flowed into the lagoons and marshes, which completely disappeared', forcing the animals to live in harmony once more.[14] Setting aside the metaphors, it is again plausible to interpret this as an effect of the rising postglacial sea level overtopping the lip, 50m (165ft) below the present sea level, at the mouth of Spencer Gulf, and rapidly flooding its interior, which had the fortuitous outcome of forcibly separating warring tribes.

Flinders knew that the land on either side of Spencer Gulf was inhabited, for he saw smoke from many fires, some even large enough to navigate by, and heard dogs howling at night and possibly even the sounds of human voices. The situation was quite different when he left the area to continue his circumnavigation. His next landing took place on Kangaroo Island, a 4,400km² (1,700mi²) island some 30–40km (19–25 miles) off the mainland where in contrast: 'neither smokes, nor other marks of inhabitants' could he see:

> *There was little doubt ... that this extensive piece of land was separated from the continent; for the extraordinary tameness of the kanguroos [sic] and the presence of seals upon the shore, concurred with the absence of all traces of men to show that it was not inhabited.*[15]

Kangaroo Island in 1802 was indeed uninhabited, since mainland Aboriginal peoples did not have watercraft that were able to successfully negotiate the often turbulent water passages separating it from the mainland. They probably also had little inclination to try to reach it, because to the local Ramindjeri people Kangaroo Island was known as Karta – the Land of the Dead – the place to which spirits of the recently deceased travelled. And yet, as has become abundantly clear from scientific studies of Kangaroo Island, it was once home to a sizeable living human population. The earliest hint of this came from studies in the 1930s that identified stone tools scattered around a former lagoon in the island's south, a discovery that prompted more focused archaeological interest. Several cave and rock-shelter sequences were later analysed that showed people to have been living there at least 16,000 years ago, and probably far earlier, when Kangaroo Island was attached to the Australian mainland during the lower sea levels of the last ice age.

Why, you might justifiably ask, were no people there when Flinders made landfall in 1802? To understand the reason for this, we might note that it was not just Kangaroo Island that was uninhabited at this time, but also many other smaller islands off the mainland (although not large Tasmania). The answer seems to lie in island size and its associated capacity to sustain a viable human population, particularly after sea-level rise caused ocean distances separating particular islands from mainland shores to become too long to be regularly or even easily traversed. At such a point, the islanders would have to have made a difficult decision: decamp to the mainland while they still could, or stay on their island without the certainty that they would be able to retain contact with the mainland. We know nothing of the former, but we do know that some took the second option and stayed put. Whether their descendants all subsequently died on the island, perhaps unable to access sufficient food and water during a prolonged drought, or unsuccessfully attempted to escape its confines, is unknown. However, what is certain from the archaeological evidence is that a once-thriving population on Kangaroo Island had completely disappeared by about 2,000 years ago, leaving 'a classic mystery story'.[16]

The numerous Aboriginal stories about the drowning of the shortest land connection – named Backstairs Passage – between Kangaroo Island and the Australian mainland (shown in Figure 3.3) all begin with 'a tall and powerful man' named Ngurunduri (sometimes Ngurunderi or Nurunderi), an ancestor to many Aboriginal groups in the region who is said to have travelled down the Murray River valley to the coast.[17] Ngurunduri had two wives and, in most versions of the story, gave them cause to run away from him.[18] With vengeance in mind, he pursued them from the mouth of the Murray River to Kangjeinwal, where he could hear them bathing at King's Point.[19] When he reached King's Point he saw them at Newland's Head, and when he reached there he saw them in the distance walking along long Tankalilla Beach. His wives spotted him coming, so began to hurry, intent on reaching the sanctuary of Kangaroo Island, which at that time 'was almost connected with the mainland, and it was possible for people to walk across'.[20] Finally, the two women reached Tjirbuk and, gathering their belongings, began to walk across to Kangaroo Island. When they had got halfway Ngurunduri reached Tjirbuk and, knowing that they sought sanctuary on the island, he roared, '*Pink'ul'uŋ'urn 'praŋukurn*' (Fall waters–you).[21]

Then the ocean rose, drowning Backstairs Passage, and 'churned', drowning the two women and carrying them southwards, where they were turned to stone, forming the Meralang (islands). The larger of these three islands is the older woman, the next in size the other, and the smallest the basket she threw off in a vain attempt to survive. In most accounts, the water rise is portrayed as catastrophic, a 'terrible flood' in an 1873 version,[22] 'tempestuous waves' in another.[23] This detail may well have been added as a plausible explanation of how Backstairs Passage – formerly remembered as traversable – became inundated. But its drowning is the key point of all these stories, many of which tell that after Ngurunduri's wives were literally petrified, he

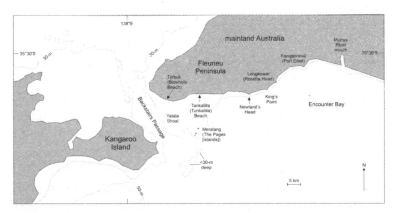

Figure 3.3 *Kangaroo Island, Backstairs Passage and the Fleurieu Peninsula.*

travelled to Kangaroo Island ... but not on foot. It seems that because the crossing was now permanently underwater, Ngurunduri dived into the ocean and swam to Kangaroo Island, from which he eventually ascended to Waieruwar – the sky.[24]

Any route across Backstairs Passage today involves ocean depths of more than 30m (100ft). It might have been possible to cross by walking, wading and perhaps a few short swims when the sea level was 32m (105ft) lower than it is today; the Yatala Shoal, which is less than 10m (33ft) deep today, may have been a critical staging post. Yet it would take a sea level that was 35m (115ft) lower than it is today for a land bridge to be fully emergent here. So the most parsimonious interpretation is that the wealth of Aboriginal stories about the crossing of Backstairs Passage dates from a time in the past when the sea level was this much lower.

Much of the south coast of Australia comprises massive bays formed by an alternation between rocky headlands and long, crescentic sweeps of windswept sandy dune-backed beaches that often impound – except during occasional floods – some of the rivers that sluggishly drain this part of the continent. One such bay is **MacDonnell Bay**, a place where the land is said once to have extended much further out to sea. There is only one version of the Aboriginal story that describes and explains what happened. It belongs to the Bunganditj people and recounts how this land then appeared: 'a splendid forest of evergreen trees, including a wattle out of which oozed a profusion of delicious gum, and a rich carpet of beautiful flowers and grass'.[25] Unfortunately, the forest was owned by a man who detested trespassers, so when one day he found a woman up a wattle tree stealing gum,[26] he told her he would drown her for her crime:

> Filled with rage, he seated himself on the grass, extended his right leg towards Cape Northumberland (Kinneang) and his left towards Green Point, raised his arms above his head, and in a giant voice called upon the sea to come and drown the woman. The sea advanced, covered his beautiful land, and destroyed the

offending woman. It returned no more to its former bed, and thus formed the present coast of MacDonnell Bay.

If, again, this strikes you as a bit far-fetched, then consider that a bare-bones story about the effects of the rising sea level drowning this coastline may not have been considered sufficiently memorable for oral transmission across many generations. Yet a story with a moral would last much longer, for then storytellers would understand its importance for proper behaviour while listeners learnt that disobedience is inevitably punished.

There is no mistaking this story for a flood story or even the recollection of a giant wave (like a tsunami), for the sea 'returned no more to its former bed'. This detail, so common in the Aboriginal coastal drowning stories outlined in this chapter, is invariably absent from the flood stories that are part of the oral traditions of coastal cultures in most other parts of the world.

The story from MacDonnell Bay is vague about exactly how much land might have been lost. So if we suppose that it refers to a time when the sea level was 15m (50ft) lower than it is today, then the 'splendid forest' might have covered an area of land extending 50–70m (165–230ft) offshore. Alternatively, had the sea level been 50m (165ft) lower at the time of the story, the land might have extended several hundred metres out to sea, perhaps indeed 'as far as the eye could carry'.

The next stop on our journey around the coast is at the doorstep of Melbourne, one of Australia's largest cities, which is built around the mouth of the Yarra River. Melbourne is sited where it is because of the proximity of the 1,930km² (745mi²) almost-enclosed harbour of Port Phillip Bay, which opens today to the Southern Ocean only through a narrow 2km (1¼ mile) wide passage (The Rip). In the heyday of cross-ocean trade, long before Australian cities were connected to one another by road, the shelter afforded to ships by such an extraordinary place was valued hugely (Figure 3.4).

The skeleton of **Port Phillip Bay** developed millions of years ago, when land was uplifted along the seaward side of what was then a sizeable coastal inlet, effectively blocking it off from the sea; on the western side the Bellarine Peninsula is a horst, an upthrust block, while along the southern and eastern sides successive phases of movement along the Selwyn Fault produced the Nepean and Mornington Peninsulas respectively. Enclosed and with several rivers debouching into it, over time Port Phillip Bay has become choked with sediments, mostly washed down by rivers but supplemented with wind-blown material and, particularly in its southern parts, marine sands washed in by waves. Yet when the sea level dropped during the last ice age, Port Phillip Bay – which is no deeper today than 25m (82ft) below sea level – dried out. The rivers cut channels through the exposed sediments and found their way to the sea. When the sea level rose once more after the last ice age ended, some of those former outlets became plugged by massive amounts of sediment, so that even when the ocean surface had risen above the level of the floor of Port Phillip Bay, it remained dry. Later, of course, the sea forced its way through at least one of those sediment plugs – that is where The Rip is today – and flooded the Bay, but it is also possible that subsequently it became blocked again and Port Phillip Bay dried up once more, even though the ocean surface was higher than its floor.[27]

Compared to many other accounts, most Aboriginal stories describing how Port Phillip Bay appeared when it was dry and how it was later inundated are matter of fact and apparently unembellished, suggesting that they may recall exactly such a comparatively recent drying-up event. Alternatively, such stories might recall the time when the rising postglacial sea level did drown the Bay, a memory that was reinforced by a more recent such event.

In 1841, Georgiana McCrae, daughter of the 5th Duke of Gordon, moved from her native Scotland to Port Phillip (as Melbourne was then known), and four years later built a house in the shadow of a granite monolith – Arthur's Seat – at the southern end of the Mornington Peninsula. By displaying

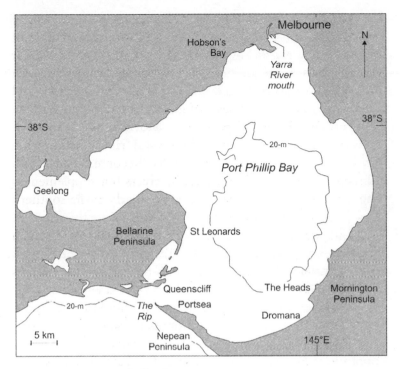

Figure 3.4 *Port Phillip Bay.*

'friendly curiosity and an unusual willingness to understand tribal customs' of the resident Bunurong Aboriginal people, McCrae – an inveterate diarist – was able to record a wealth of information, peppered with verbatim quotes, about their way of life and their stories about Port Phillip Bay.[28] She wrote:

> *The following is an [Aboriginal] account … of the formation of Port Phillip Bay: 'Plenty long ago … men could cross, dry-foot, from our side of the bay [in the east] to Geelong [in the west].' They described a hurricane – trees bending to and fro – then the earth sank, and the sea rushed in through the Heads, till the void places became broad and deep, as they are today.*[29]

A decade or so later, on 9 November 1858, in his submission to a Committee of the Legislative Council enquiring into the condition of Aboriginal peoples, one William Hull recalled that various Aboriginal groups 'say that their progenitors recollected when Hobson's [Port Phillip] Bay was a kangaroo ground – they say "plenty catch kangaroo and plenty catch opossum there"', a condition that could have obtained only if the Bay was dry land. Hull went on to add that his Aboriginal informants had told him that 'the river Yarra once went out [to sea] at the Heads, but that the sea broke in, and that the Bay became what it is [today]'.[30]

Another historical recollection was collected in the 1950s and recalled that in pre-European times, the Aboriginal people at Dromana were accustomed to cross dry Port Phillip Bay to hunt at Portsea and Queenscliff; in doing so, they had to 'walk a little, swim a little'.[31] And then there is a mythical story handed down among the Kulin Aboriginal people which explains that

> …Port Phillip was once dry land and the Kulin were in the habit of hunting kangaroos and emus there. One day the men were away hunting and the women had gone off collecting roots and yams, while some young boys, who had been left behind, were playing in the camp. They were hurling toy spears at each other, just like their fathers did. In the camp there were some wooden troughs full of water, and one of the spears upset one of these … this was no ordinary bucket, but a magic one, and it held a tremendous amount of water, which came rolling down engulfing all the land.[32]

In order for Port Phillip Bay to become dry, the sea level outside The Rip would have to be at least 9–12m (30–40ft) lower than it is today.

East from Port Phillip Bay and Melbourne lies the coast of Gippsland, the closest part of the mainland to Tasmania, today Australia's largest offshore island. Gippsland is a low-lying region also characterised by alternating rocky headlands and long sandy beaches. It lies along the northern side of Bass Strait, which for much of the time people have lived in Australia was dry land – the Bassian Land Bridge – that allowed people and animals to move freely between the two. The postglacial sea-level rise is known to have progressively drowned this land bridge, the last tenuous connection being severed about 14,000 years ago, thereafter consigning the people of Tasmania (and islands in the Bass Strait) to progressively increasing isolation from their mainland cousins.[33]

Along low-lying coasts like that of **Gippsland**, the effects of the postglacial sea-level rise would have been particularly dramatic, so it is unsurprising that there is an Aboriginal story recalling them. The story belongs to the Kurnai people and recalls that

> ...long ago there was land to the south of Gippsland where there is now sea, and that at that time some children of the Kurnai, who inhabited the land, in playing about found a turndun [a musical instrument], which they took home to the camp and showed to the women. 'Immediately,' it is said, 'the earth crumbled away, and it was all water, and the Kurnai were drowned.'[34]

Unwittingly the Kurnai children had broken an important taboo, for the *turndun* was only for men's use; its discovery by children and handling by women invited retribution. Like enduring oral traditions in many cultures, the use of retribution as an explanation of why things happen is common in Aboriginal cultures, probably because adults felt an ethical imperative to pass on morality stories to their young that would not have been applicable to neutral ones.

Since there is just the one story from Gippsland and since it is so vague about where the coast once stood, we can only speculate about this and the associated level of the sea. If we assume that the story recalls a time when the shoreline lay some 50m (165ft) seaward of its present position, then the sea level might have been around 20m (65ft) lower. If the Gippsland shoreline was 100m (330ft) further out, then the sea level would have been closer to 50m lower than it is today. This information is key to estimating minimum ages for such stories, something that is explained at the end of Chapter 4.

From Gippsland, we head up the east coast of Australia to Botany Bay, the place where Captain James Cook first landed on 29 April 1770, which is now part of the sprawling city of Sydney. The description of Botany Bay by Joseph Banks, naturalist on Cook's voyage, was so glowing that 18 years later it was the place where Governor Arthur Phillip was instructed to establish Britain's first penal colony in Australia. It did not take Phillip long to realise its unsuitability for this purpose – its marshy foreshore, and the lack of fresh water and a secure anchorage – so he swiftly shifted the nascent penal colony to Sydney Cove, one of several bays that eventually became the settlement of Port Jackson. We are left to wonder why initial reports about Botany Bay were so wrong. Consider the experience of Captain Watkin Tench, whose books about the first four years of the Port Jackson settlement are among the most informative from that era. Criticising Banks and the other 'discoverers of Botany Bay' for noting that its fringes were 'some of the finest meadows in the world', Tench noted that 'these meadows, instead of grass, are covered with high coarse rushes, growing in a rotten spungy [sic] bog, into which we were plunged knee-deep at every step'.[35]

Yet despite such an unpromising description, **Botany Bay** has a Aboriginal story, owned by the Dharawal people, that talks about how the Bay formed as a result of sea-level rise. It tells of a time long ago when the Bay was mostly dry land, the floodplain of the Kai'eemah (today's Georges River). So grateful were the Dharawal for having such a fine place to live that one day the decision was made to travel into the hinterland to give thanks to the Creator Spirit for this. The younger members of the community were not happy at the prospect of travelling into such 'rough land', so in the end only the 'knowledge holders' went, leaving a warrior named Kai'mia behind to look after the others. After several days, huge waves washed into the mouth of the Kai'mia, 'destroying much of the swampland … used for food gathering'. Fleeing inland, Kai'mia and the youngsters were pursued by giant waves, eventually taking refuge in a cave. Kai'mia tried persuading his charges that they should give thanks to the Creator Spirit for having saved them, but one retorted that they now had nothing to be thankful for – 'the waves have taken away what we had'. Then the cave collapsed, burying all its occupants except Kai'mia, who crawled away to die, his trail of blood marked by the flowering of the Gymea Lily – a plant with blood-red tips. When the knowledge holders returned, they followed the trail of the Gymea Lily back to the cave, but were too late to rescue those buried inside. Sadly, they 'returned to their homeland to find that what they had known was no longer. Instead of the swamps, there was a great bay, and where the Kai'eemah had met the sea there was [sic] high mountains of sand'.[36]

Today, Botany Bay has been transformed beyond recognition – Sydney's airport is located on its shore – so the most reliable guide to its original form is the chart made by Captain Cook in 1770. This shows that the entrance to Botany Bay was then 5–9 fathoms deep, a range of 9–16m (30–50ft), the depth at which the ocean surface would need to be for Botany Bay to be dry land – as the Dharawal recall it once was.

The bay just south of Botany Bay is **Bate Bay**, where 'Mister', perhaps one of the last of the Gunnamatta Aboriginal people who once occupied the area, passed on his reminiscences in the 1920s. One of his stories was about how 'in the early days the sea was a lot further out, and his people used to gather ochre there', at a place he identified as being about 4km (2½ miles) east (seawards) of Jibbon Headland.[37] If the '4km' is taken literally, the modern ocean depth here is about 50m (165ft); more likely, the stated measure was unintentionally vague, perhaps much less than 4km, so the depth to which this memory refers is more likely 5–20m (16–65ft) below today's ocean surface.

Further north from here, off the mouth of the Brisbane River, lie two elongate islands – Moreton and North Stradbroke in English, Moorgumpin and Minjerribah respectively pre-colonisation. **Moreton** and **North Stradbroke Islands** are formed predominantly from sand, typically a 'bedrock' of hard indurated dune sand overlain by a veneer of younger unconsolidated sand, all of it deposited along the shores of these islands by waves that had carried it from the south. During the last ice age, when the sea level was lower than it is today, these islands were joined to the Australian mainland. Therefore at the time of the earliest evidence for their human occupation – about 20,000 years ago at Wallen Wallen – no cross-sea journeys were required. Yet by about 7,000 years ago, the sea level had risen so much that both islands were effectively cut off from the mainland, and their inhabitants thereafter developed traits distinct from their mainland counterparts.

The Aboriginal people who live on Moreton Island are known as the Nughi, those in the northern parts of North Stradbroke as the Noonukul, and both groups are part of the Quandamooka nation. Stories about times in the past, when the geography of these islands was quite different from what it is today, include both mythical and factual varieties. The sole known representative of the former comes from a time when the two islands were one. It involves a Noonukul man named Merripool who had the power to control the winds, something he carried around in a magic bailer shell.[38] Envious of such power, the Nughis plotted to steal this shell, but Merripool was informed of their plan and

> ... *called the four winds to come to him* ... *He called for many days and nights and his voice became louder and louder* ... *[until] the winds blew so hard that they caused the water to cut Moorgumpin [Moreton] away from Minjerribah [North Stradbroke].*[39]

In a detail that perhaps expresses the ecological consequences of separation, the story continues by saying that when the animals on Moreton Island realised what was about to happen, they all moved to North Stradbroke Island, leaving the Nughi people on Moreton Island solely dependent on the sea for their food.

Evidence of a factual memory of a time when Moreton and North Stradbroke Islands were apparently not as far apart as they are today comes from the account of Mary Ann, an Aboriginal woman who was interviewed in 1907, when she was 'very old', and who remembered that when she was young 'a small island' existed between the two on which 'bopple bopple' trees grew.[40] She also recalled that at the time the Noonukul elders could themselves recall a time when the people of North Stradbroke Island might converse 'quite easily' across the gap with the Nughis on Moreton Island.[41]

On 17 May 1770, Captain James Cook gave the world the first written account of these islands. He named the bay in which they are found Moreton Bay, noting in his journal that the land forming the islands he 'could but just see it from the top mast head'. Interestingly, he did not record there being two islands here and it is possible that at this time the islands were joined, forming a continuous strip of land.[42]

Sand islands are notoriously changeable in form and many periodically merge or become severed from their neighbours because of changes in the amount of sand being supplied to their shores by waves – rather than any change in the sea level. So while the stories about Moreton and North Stradbroke Islands are certainly intriguing, they do not necessarily require – as do most other stories related in this chapter – the sea level to have been lower than it is today. That said, we know that the islands were connected when the sea level was lower 9,000 years or so ago, so perhaps the possibility that these stories are recollections of that time should not be too hurriedly dismissed.[43]

In a similar vein, the next places along the Australian coast for which there is an Aboriginal tradition of a time when islands were not islands are **Hinchinbrook** and **Palm Islands**, off the central Queensland coast. Hinchinbrook and Palm Islands are quite different, with the former being much larger than the latter and much closer to the mainland. Palm Island today lies about 25km (16 miles) offshore. The people of this area have a story about a distant ancestor of theirs named Girugar, who travelled across the area giving key places the names by which they are still known and – most critically for our purposes – walking across to Hinchinbrook and Palm Islands (and nearby islands) without getting his feet wet.[44] The only way that this could be accomplished would be if the ocean surface were at least 22m (72ft) lower than it is currently.

In his 1980 opus *The Languages of Australia*, Robert Dixon, an indefatigable recorder and analyst of Aboriginal languages, noted how 'many tribes along the south-eastern and eastern coasts [of Australia] have stories recounting how the shoreline was once some miles further out; that it was – on the north-east coast – where the [Great] barrier reef now stands'.[45] Inscribed on the World Heritage List, the Great Barrier Reef – the only living thing on Earth that is visible from outer space – lies off the east coast of tropical Australia. Around its southern extremity its outer edge is some 200km (125 miles) offshore, but farther north, in the area around Cairns, this is a mere 40km (25 miles) seawards of the modern coast. As shown in Figure 3.5, the ocean floor between the modern shoreline and the reef edge is comparatively shallow, with patches of reef separated by deeper water passages. Drop the sea level a few tens of metres and these patch reefs become hills and the intervening passages become valleys, a gently undulating landscape that would have terminated during the last ice age 20,000 years ago in sheer cliffs – today's reef edge – around the base of which the waves would have crashed resoundingly.

Cairns is built on the traditional lands of the Yidinjdji people, who also claim title over much of the submerged land extending to the edge of the Great Barrier Reef, for many parts of which they have names and about which they know stories. The most widely reported story about coastal drowning in this area was first written down between 1892 and 1909 by the Reverend Ernest Gribble, when he was in charge of the Yarrabah mission station. Gribble wrote of the tradition concerning a man named Goonyah, who by catching a forbidden species of fish angered the 'Great Spirit' Balore. In an attempt to drown Goonyah and his family, Balore caused the sea to rise. The humans fled to the mountaintops, where they heated large stones and rolled them down the slopes, something that checked the advance of the rising waters; 'the sea, however, never returned to its original limits'.[46] A later version of this story, collected in the vernacular (Yidin) language by Robert Dixon, also included the detail that the Yidinjdji had tried in vain to stop the unwanted rise of the ocean across the lands on which they had long depended for sustenance.[47]

Another story collected from the Djabuganjdji (Tjabogai-tjanji) people in the late 1920s from Double Island, just north of Cairns, recalls a time 'when the coral reef was all scrubland' and a blue-tongued lizard, of the kind common in eastern Australia, 'travelled to the edge of the deep dark waters and caused the sea to bubble up till it covered the reef and arrived at its present position'.[48]

Echoes of the time long ago, when the area between the modern coastline around Cairns and the edge of the barrier reef were 40km (25 miles) or so away, are also found in the

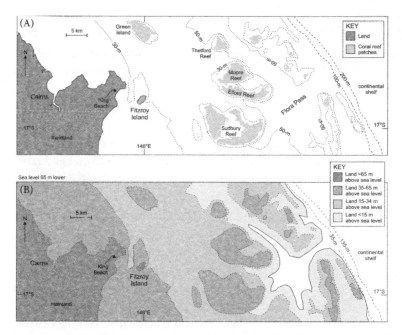

Figure 3.5 *Cairns and the Great Barrier Reef, today (A) and during the last ice age (B).*

names of places in the Yidin language.[49] Key is that the Yidin word for island is *daruway*, which also means 'small hill'. The Aboriginal name for Fitzroy Island is *gabar*, which means 'lower arm', a reference to the time when it was a mainland promontory enclosing a river valley. There was once an island named Mudaga halfway between Fitzroy Island and King Beach, named for the pencil cedar trees (*Polyscias murrayi*) that grew there when it was emergent. Finally, the Yidinjdji recall a time when Green Island was formerly four times larger than it is today, something that would be expected were the sea level lower.

For Green Island to have been four times larger than it is today, the sea level might need to be 5–10m (16–33ft) lower, but Fitzroy Island could not have been the 'lower arm' of a mainland promontory unless the sea level was at least 30m (100ft) lower. Yet if we accept the stories about the Great Barrier Reef being dry or even 'scrubland', the sea level would need to have been 40–50m (131–165ft) lower. And if indeed the Aboriginal people of the area knew a time when the shoreline was where [the edge of] 'the barrier reef now stands', then this would have to have been when the sea level was a staggering 65m (210ft) lower than it is at present.

Our next stop along the Australian coast is on the shores of its largest indentation, the Gulf of Carpentaria. During the last ice age, when the sea level was more than 120m (400ft) below today's level, this part of Australia was connected to New Guinea to the north. At the centre of the land bridge lay a great freshwater lake – Lake Carpentaria – around which the people and animals of the area flocked.[50] The rising sea level after the ice age ended saw the progressive inundation of this land bridge, the eventual conversion of the freshwater lake to the saltwater gulf, and the appearance of islands within the Torres Strait that today separates Australia from Papua New Guinea.[51] The progressive submergence of this well-populated land bridge would of course have driven many groups of people outwards to seek new land from which they might subsist.

The languages spoken by Aboriginal groups in most parts of Australia around the time of its European colonisation all belong to the same language family (Pama-Nyungan), which is thought to have spread across most parts of the continent some time within the past 3,000 years or more. What is key for our purposes is that this dispersal appears to have originated somewhere around the now-submerged land bridge between Papua New Guinea and Australia, and it is plausible to suppose that its outward spread from here was driven by sea-level rise. Some corroboration of the precise and quantifiable linguistic argument may be found in Aboriginal stories about 'lost lands', now-submerged homelands from which ancestral mainland populations came. The geographical details about where these homelands were located are typically vague, but one that is told by the Yolngu people of eastern Arnhem Land, who occupy part of the western shores of the Gulf of Carpentaria, states that the ancestral land named Bralgu was located 'somewhere in the Gulf of Carpentaria', perhaps therefore a direct memory of the translocation of people from part of the submerging land bridge.[52]

Our focus is on islands lying off the coast within the modern **Gulf of Carpentaria**, principally the Wellesley Islands (Figure 3.6), which would have formed only after the sea level rose across the former mainland fringe. The earliest written version of the oral tradition was produced by Dick Roughsey of the Lardil people in the early 1970s, who explained that:

In the beginning, our home islands, now called the North Wellesleys, were not islands at all, but part of a peninsula running out from the mainland. Geologists … thought that the peninsula might have been divided into islands by a big flood which took place about 12,000 years ago. But our people say that the channels were caused by Garnguur, a sea-gull woman who dragged a big walpa or raft, back and forth across the peninsula.[53]

The Lardil also remember that the Balumbanda, one of their clan groups, originally came from the west along that peninsula before it was 'cut up into islands'.

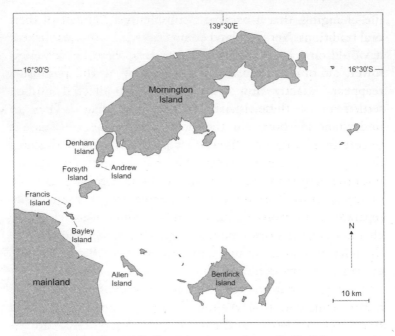

Figure 3.6 *The Wellesley Islands, southern Gulf of Carpentaria.*

Later research focused on Lardil stories dated to more than 5,000 years ago, about how channels formed that isolated the Wellesley Islands from the mainland.[54] A fuller version of the Garnguur (*kankurr*) story told by Roughsey explains that the seagull woman sought to punish her brother Crane for failing to look after her child, so she dragged her large raft back and forth across the land that connected Forsyth Island to Francis Island until the deep-water channel (more than 10m/33ft deep) that exists there today was formed. The sweet-potato woman, Puri, who was fleeing from Crane, is said to have created the passage (named *purikal*) between Forsyth and Andrew Islands. The narrow channel between Denham and Mornington Islands is said to have been created by a shark panicked by lightning.

The sheer number of channel-cutting stories here suggests that the gradual submergence of the land connection between the Wellesley Islands and the mainland was witnessed by people who found the process so disruptive, so literally life-changing, that it became a conspicuous element of their oral traditions. Yet for a land connection to be re-established, it would only be necessary for the sea level to be 5m (16ft) lower, although a 10m (33ft) fall would see the peninsula reappear. Exactly how sea-level changes affected human settlement of these islands – something that is key to understanding how old the stories might be – is as yet uncertain. While it is likely that people occupied the area during the last ice age when the 'islands' were contiguous with the mainland, we cannot be sure whether people stayed on the embryonic islands as the rising sea level was starting to separate them from the mainland – which is what some channel-cutting stories suggest – or whether they retreated to the relative security of mainland shores at this time. There is evidence favouring both views.[55]

Moving west from the Gulf of Carpentaria we come to the north-facing coast of Australia, much of which is called Arnhem Land. Perhaps because it is further north than almost every other part of the continent, it has a wealth of Aboriginal traditions and stories that are likely to be some of the earliest in Australia. The four groups of drowning stories in this area all involve offshore islands – Elcho, the Goulburn group, Cape Don and the Tiwi (Bathurst and Melville) Islands. It is perhaps no coincidence that such stories are found predominantly along the coast of Australia where the continental shelf exposed during the last ice age was widest – and where, consequently, the subsequent rise of the sea level was most noticeable. One authority estimated that during this period 'on the gently sloping northern plains [of Australia] the sea inundated five kilometres of land annually'.[56] Not only is this kind of change blatant, but its effects on the way people lived, what they could eat and with whom they had to compete for food and territory undoubtedly left enduring marks on the contemporary evolution of Aboriginal societies in this part of Australia.

Elcho Island is separated from the mainland today by a narrow strait that would be passable, albeit circuitously, were the sea level 5m (16ft) lower than it is today, and fully emergent

if the sea level were 10m (33ft) lower. Two extant stories
about such a time are known, one about an Elcho resident
named Djankawu who tripped while walking along the
beach, accidentally thrusting his walking stick into the sand
and 'causing the sea to rush in'.[57] The other tells of the
Ancestresses, magical beings who, whenever they needed to
cross to Elcho, made a sandbar that vanished once they had
completed their crossing.[58]

Further west off the Arnhem Land coast lie the **Goulburn
Islands**, comprising South Goulburn (Warruwi) and North
Goulburn (Weyra). One story, which is perhaps an echo of
the time when Warruwi was joined to the adjacent mainland,
concerns a man named Gundamen who wanted to cross from
the latter to the island. Afraid of the deep water, he was
helped by a woman with magic powers who called out '*mubin,
mubin, murbin!*', which caused a land bridge to emerge just
long enough for him to cross.[59] Another story is more detailed
and recalls a time when only a small creek named Mandurl-
mandurl separated Warruwi from Weyra. The creek teemed
with fish, and the local people would set their nets (*yalawoi*)
in it overnight and pull them up full in the mornings. For all
this, a man named Gurragag rarely succeeded in catching any
fish in Mandurl-mandurl, so one day he cut down a huge
paperbark tree (*waral*). It fell into the creek with such an
enormous splash that seawater poured in, creating the ocean
gap between the two islands that is today some 7km (4½
miles) wide.[60] If such stories recall a time when these islands
were in fact closer to the mainland and to each other, then
the sea level would have to have been 17–20m (56–65ft) lower
than it is today.

The people of coastal Arnhem Land 'possess names for and
maintain intimate knowledge of places far out at sea that are
known to have been above sea level 10,000 years ago', an
extraordinary yet undeniable fact that has become the basis
for a large-scale marine-protection strategy for this part of
the Arafura Sea.[61] For example, the inhabitants of the
Goulburn Islands know of two submerged islands – Lingardji
and Wulurunbu – thought to be the sites of shoals about

20km (12 miles) east of Weyra that now lie in waters 9–11m (30–36ft) deep. How could these people have such intimate knowledge of a submerged landscape unless their ancestors had once walked across it, witnessed its submergence and consigned their memories of it to posterity?

Moving west along the coast of Arnhem Land, we come to **Cape Don**, the western extremity of the Cobourg Peninsula, which has been occupied by the Arrarrkbi peoples for tens of thousands of years. They have stories about how one group of their ancestors once occupied an offshore island named Aragaládi that is now underwater. One story tells that a sacred rock (*maar*) on the island was accidentally bumped one day, causing so much rain to fall that the island became submerged: 'children and women were swimming about … [they] had no canoe to enable them to cross over in this direction to the mainland at Djamalingi (Cape Don) … trees and ground, creatures, kangaroos, they all drowned when the sea covered them'.[62] This story perhaps combines memories of large waves like tsunamis and their effects on coastal peoples in this area with the story of an island being submerged, a common situation with similar oral traditions in the island groups of the Western Pacific Ocean.[63]

There is another story about the disappearance of Aragaládi that is more entwined with other oral traditions of Arnhem Land Aboriginal peoples, particularly that of the Rainbow Serpent. For this story has Aragaládi as part of a giant snake, one of its coils conveniently (and intentionally) raised above the ocean for people to live upon. But when its inhabitants knocked the *maar*, the snake drew its coil underwater, swallowing all the people: 'she made the place deep with sea water. Those first people became rocks. Nobody goes to Aragaládi now.'[64]

The Mayali people, who now live inland east of the Alligator River (see Figure 3.7), also have a tradition that the first people in this part of Australia – the Nayuyungi – came from the sea between the Cobourg Peninsula and the Goulburn Islands. This is also possibly a distant memory of an inhabited island that was submerged by postglacial sea-level rise, compelling its inhabitants to relocate to the mainland.

The key point here is that along the Arnhem Land coast, likely to be the longest continuously inhabited part of Australia, as well as the one with the broadest continental shelf to be drowned by postglacial sea-level rise, there are numerous extant stories (and probably many more forever lost) that tell of islands off the coast that were once inhabited but are now submerged. This is just what would be expected in such a geographical situation that experienced rapid sea-level rise over several millennia, yet it is impossible to know the minimum depth that the sea level would have to have been in order for these islands to have been emergent. If we assume, without any real evidence, that Aragaládi Island was once the shoal some 150km (93 miles) east of the Cobourg Peninsula, then the sea level would have had to be 9–15m (30–50ft) lower than it is today for habitable land to be exposed there.[65]

The next stop on our tour of the Australian coast is a group of sizeable offshore islands – the Tiwi Islands – where traditions tell of the apparent isolation of their populations as the sea level gradually rose. Better known as **Bathurst** and **Melville Islands** (Figure 3.7), they lie a minimum of 25km

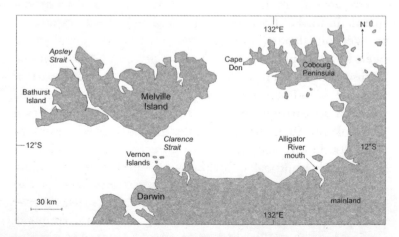

Figure 3.7 *Location of Bathurst and Melville Islands (the Tiwi Islands), Cape Don and the Cobourg Peninsula.*

(16 miles) off the Australian mainland near the northern city of Darwin. It is to the water gap known today as Clarence Strait that Tiwi stories about the apparent cutting off of these islands from the mainland refer.

The stories begin in the Tiwi 'time of darkness', before Bathurst and Melville were 'born' as islands, a time when the land 'contained no geographical features, animals or humans'.[66] Then, from within the earth – meaning from what may have been becoming the mainland (rather than the islands) – came an old blind woman named Mudangkala (or Murtankala), who crawled with her three children along what is now Clarence Strait until they reached Melville Island. Critically, water followed the group as it moved northwards: 'the flow of water continued to increase and is today known as Clarence Strait ... [Mudangkala] continued to move over the land known as Bathurst Island till finally water flowed on to form what is now known as Apsley Strait'.[67] These stories may be memories of one of the last significant human migrations from the Australian mainland to the Tiwi Islands, the references to blindness perhaps signifying the uncertain nature of Mudangkala's destination, and those to crawling indicating the difficulty of crossing a partly submerged land bridge.

For anyone in the past attempting to travel from the mainland to the Tiwi Islands without the use of watercraft, the Vernon Islands (Potinga to the Tiwi people) would have been key 'stepping stones'. One Tiwi story recalls that the Vernon Islands were once attached to the southern part of Melville Island at a place called Mandiupi, but then, 'as a result of an earthquake in the distant past', the connection was broken.[68] Such a story illustrates the issue of the plausibility of oral traditions well. In order to expect a tradition – like the instinctively improbable 'breaking-off' of islands – to be duly repeated through future generations, it may need some explanation. Not associating the story with a sea-level rise, perhaps the storyteller added the detail about an earthquake simply to make it credible. For an earthquake could not possibly have caused the Vernon Islands to break off from

Melville, but a rising sea level would have swamped the connection, giving the same apparent result.

Other aspects of Tiwi society, from its language to its uncommon matrilineal society, make it distinct from mainland Aboriginal societies, something that is ascribable to physical and cultural isolation imposed by postglacial sea-level rise.[69] In addition, there are various aspects of Tiwi Island terrestrial ecosystems that also point to the effects of isolation.[70] So for how many years have the islands been isolated, and how long is it since Mudangkala made her last desperate crossing of Clarence Strait? We have to wait until the end of Chapter 4 for a precise answer to these questions, but for now it seems that isolation would have occurred when the ocean around the Tiwi Islands was at least 12m (40ft) lower than it is today, a time in the past when – through a combination of walking, wading and short swims – it would probably have been possible to reach them from the mainland. When the sea level was 20m (65ft) lower, it is likely that someone could have crossed without getting their feet too wet, although the likely route would have been tiresomely circuitous.[71]

Moving further west into north-west Australia, it is surprising, given this region's demonstrably long history of Aboriginal occupation and its extraordinary legacy in rock art (throughout the Kimberley and Pilbara regions in particular), that there are not more stories about coastal drowning here than there are. In fact, there is just the one group of stories – about Brue Reef (discussed below) – but there are also two intriguing studies that demonstrate the massive impact which postglacial sea-level rise had on the Aboriginal inhabitants of this region.

The first of these studies concerns the baobab tree (*Adansonia gregorii*), which is native to north-west Australia as well as parts of Africa. In Australia it is usually called the bottle tree, and for thousands of years it has been prized by savannah-dwelling Aboriginal groups for food, medicine and shelter from the sweltering conditions in which it thrives, typically from the coast to the desert fringes. Parallel lines of research into baobab gene flows and into the routes along which the various words for baobab among Aboriginal groups passed allow it to be plausibly shown that these trees, once confined to the (now-submerged) continental shelf of north-west Australia, were probably carried inland and intentionally dispersed as the sea level rose in this area.[72] The clever idea behind this scenario is that the oldest (genetically less diverse) types of baobab are found along the coast, where the people also have the fewest words for the tree (and its component parts). Further inland there is greater genetic baobab diversity and a far greater range of names for the tree. These two observations suggest that people living with baobabs along the now-drowned continental shelf, recognising their great value, carried their fruit pods with them as they were gradually forced inland by the rising waters. In addition to deliberately planting the baobab inland, they also introduced it to inland Aboriginal groups that, quickly convinced of its value, dispersed it even more widely.

The second study used Aboriginal rock art to plot the course of postglacial sea-level rise in the Dampier Archipelago (Murujuga). The earliest rock art here dates from the time during the last ice age when the sea level was so low that today's islands were part of a mountain range some 160km (100 miles) inland from the coast. With the rise of the postglacial sea level, the area was gradually transformed – from inland to coastal to offshore islands – but the people stayed put, adapting their livelihoods to the changes that the sea level (and climate) imposed on them. They recorded these changes in their rock art that is fortuitously etched through a veneer of rock varnish into the weathered surfaces of the gabbro and granophyre boulders that are scattered about the Murujuga landscape in vast numbers. They look like scree but have in fact weathered out of the underlying bedrock in the places where they are found now, unmoved for perhaps millions of years. Thus, the rock art dating from when the area was far inland shows examples of inland fauna like macropods (kangaroos and wallabies), but in the rock art from the time the shoreline had reached the area, engravings of fish and other sea creatures are dominant.[73]

Stories about **Brue Reef** are well known among the Bardi and Jawi peoples of the Kimberley coast.[74] Lying about 50km (30 miles) off the mainland and uncommonly isolated, Brue Reef is said once to have been a habitable island named Juljinabur; today it is awash at high tide. The gist of the various stories is that Juljinabur was inhabited by a person called Jul and his greedy kinfolk (*munjanggid*), who had become cannibals. One day, in fruitless pursuit of a turtle, a Jawi family from Tallon Island (Jalan), 100km (60 miles) to the south, found themselves drifting in a swift current (*lu*) towards Juljinabur. To save the Jawi family from being eaten, Jul hid the people on the island for a few days, and when his kinsfolk's attention was diverted, sent them back to Jalan in a double-hulled canoe (*inbargunu*). Unfortunately, the *munjanggid* got wind of what had happened, and there was a big fight that precipitated the sinking of the island and its conversion to the inter-tidal reef it is today.

For a community sufficiently large to be organised and competitive to have lived on Brue Reef, this 12ha (30 acre) island would have had to be at least 4m (13ft) above the ocean surface. Such a vision becomes altogether more believable if we suppose that the sea level at the time Jul and his predatory kin lived was perhaps 10m (33ft) lower than it is today.

We now move down the coast of Western Australia to the city of Perth, founded in 1829 on the banks of the Swan River, the traditional land of the Noongar people. Off its mouth lie three islands – **Rottnest**, **Carnac** and **Garden** – that are said once to have been connected to the mainland (Figure 3.8).

There appears to be just one extant version of this story, although it has been repeated numerous times in different contexts. It is a good example of the narrative type of story, and was in fact recorded merely as part of an appendix in George Fletcher Moore's 1884 *Diary of Ten Years Eventful Life of an Early Settler in Western Australia*, which – notwithstanding

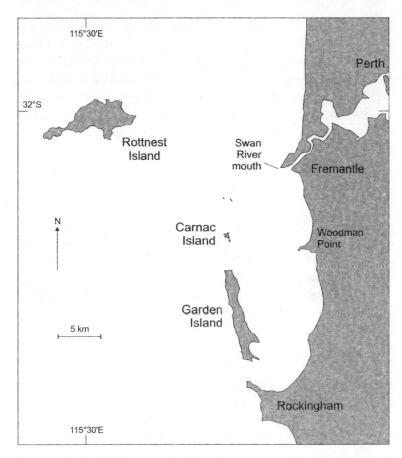

Figure 3.8 *Off the mouth of the Swan River, Western Australia, lie the islands of Rottnest, Carnac and Garden.*

its gung-ho title – amply demonstrates the author's sympathetic understanding of Aboriginal culture. The story goes

> ... *that Rottnest, Carnac and Garden Island, once formed part of the mainland, and that the intervening ground was thickly covered with trees; which took fire in some unaccountable way, and burned with such intensity that the ground split asunder with a great noise, and the sea rushed in between, cutting off these islands from the mainland.*[75]

It is probable that Moore, 'a religious man of strong convictions' according to his entry in the *Australian Dictionary of Biography*, spiced up the original narrative; the image of the 'ground splitting asunder with a great noise' has biblical resonance and is probably not an authentic detail of the Aboriginal story. Yet the bare fact of these islands – Rottnest is now some 20km (12 miles) offshore – having once been part of the mainland is something that is consistent with many other stories discussed in this chapter for other parts of the Australian coastline. So how might it ever have been so?

As for Kangaroo Island and several other islands off the Australian mainland, it has long been known that people lived on Rottnest (or Wadjemup) – the largest and highest of the three islands – in pre-colonial times, yet it was uninhabited at the time of the first written accounts of it. The first estimate of the date of the earliest human occupation of Rottnest was based on measurements of the ages of land-snail shells within ancient sand dunes burying artefact-bearing soils, and was calculated as being in excess of 50,000 years ago – and perhaps as long ago as an improbable 125,000 years. As is often the case, recent research has established a more probable time for the formation of the artefact-bearing soils as between 10,000 and 17,000 years ago.[76] Were this the case, then Rottnest would not have been an island but would have been attached to the mainland because of the lower sea level. The reason why Rottnest had no Aboriginal people living there at the time of European colonization of Australia is probably similar to those of its offshore islands. *Either* a residual population stayed there after the island was cut off from the mainland and died out because of a lack of resources and its inability to survive periods of extreme weather, particularly drought on limestone Rottnest, *or* the place was abandoned as it was becoming an island because its inhabitants feared a future in isolation. Justifiably so.

If we accept the detail in the story that Moore recorded about the land connection being 'thickly covered with trees', it does suggest that it was well above the high-water mark and perhaps significantly beyond the reach of sea spray. This condition would be adequately met were the ocean surface 10m (33ft) lower than it is today, although a narrower, meandering land connection between Rottnest and the mainland would also have existed when the sea level was just 5m (16ft) lower than it is today.

Around 15km (9 miles) off the south-west corner of Australia lie two prominent rocks. Named the **White-topped Rocks**, perhaps because they are plastered with seabird excrement, they are a well-acknowledged aid to navigation in these sometimes unsettled waters. There is a single Aboriginal tradition, perhaps collected as early as 1844, about these rocks that goes as follows:

> ... *in those olden days there was a large plain extending from the main land out to the White-topped Rocks, about nine miles [14km] out from Cape Chatham. On one occasion two women went far out on the plain, digging roots. One of the women was heavy with child, and the other woman had a dog with her. After a while they looked up, and saw the sea rushing towards them over the great plain. They both started running towards the high ground about Cape Chatham, but the sea soon overtook them and was up to their knees. The woman who had the dog picked it up out of the water and carried it on her shoulders. The woman who was advanced in pregnancy could not make much headway, and the other was heavily handicapped with the weight of the dog. The sea, getting deeper and deeper, soon overwhelmed them both, and they were transformed into the White-topped Rocks, in which the stout woman and the woman carrying the dog can still be seen.* [77]

The critical omission, common in giant-wave (tsunami) stories, that the sea did not subsequently retreat from the land it inundated, implies that this story may also be an authentic memory of coastal drowning in this area. If this is indeed the case, then for the sea floor between the mainland and the White-topped Rocks to have been dry, the sea level would need to have been 55–60m (180–200ft) lower than it is today.

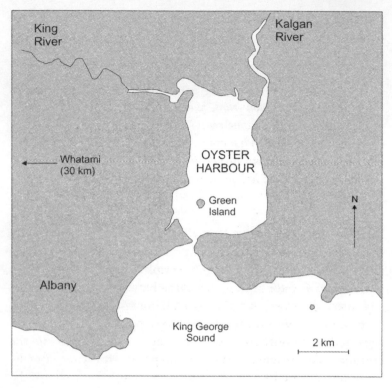

Figure 3.9 *Oyster Harbour.*

The next example, further to the north-east, comes from **Oyster Harbour**, a tidal inlet with a narrow, deep entrance – a smaller version of Port Phillip Bay (described earlier) – which is shown in Figure 3.9.

The story comes from Captain Collet Barker, an army officer who in 1829 became commander of the military settlement at King George Sound, now the town of Albany. Barker was an energetic recorder of Aboriginal place names and traditions, in which context he recorded the only known such story about the origin of Oyster Harbour.[78] It begins with a woman going into the 'bush' to search for food and calling out to her husband, who was sitting by their cooking fire. When she found a particular type of snake, a 'Quoyht' as rendered by Barker, the man was happy, but when she returned empty-handed, her stomach full of this great delicacy, he became enraged. He struck her, broke her leg, then ran away:

> *She becomes sick & dragging herself along in the line where the King's River now runs, reaches Green island, where she dies ... Her body became putrid & an easterly wind setting in is smelt by a dog at Whatami ... He follows her track & arrived at the place, commences scratching, which he continues so long that he digs a great hollow & the sea comes in & forms Oyster Harbour.*

Like the story relating to Port Phillip Bay, this one may echo events far more recent than the period of postglacial sea-level rise – but it may not. It may simply be that the narrow entrance to Oyster Harbour became blocked, as such narrow gaps often do, as a result of which the inlet dried up, a situation that might have continued for so long that local people regarded it as normal. Then, perhaps during a storm, the entrance was breached and the ocean poured in. Alternatively, it might be more like the situation at Spencer Gulf, where the rising sea level overtopped the entrance to a lowland, perhaps marshy area, and flooded it rapidly.

In the first, more conservative scenario, the sea level would not need to be any different from today's, but it would have to have been at least a few metres lower for the latter to be true; an ocean surface 4–10m (13–33ft) below its present level would adequately account for the progressive flooding of the area, and the formation of the inlet and its later submergence.

The next story is very similar, reflecting the prevalence of coastal inlets with narrow entrances along this part of the Australian coast. At **Bremer Bay**, a brackish-water lagoon named Wellstead Estuary is blocked at its seaward entrance by a barrier beach, a thick plug of sand through which water can slowly seep, but which effectively prevents the ocean from reaching the estuary and vice versa. Many such barrier beaches formed when the sea level was rising after the last ice age, with waves driving sand into massive heaps at the heads of bays like Bremer Bay. Such barriers can endure for thousands of years, and in some parts of the world they have long been populated. Parts of iconic Cape Cod in Massachusetts, USA, for example, where many houses are built on unconsolidated barrier beaches, are likely this century to experience 'widespread and catastrophic destabilization ... resulting in significant land losses and salinization of freshwater environments' as a result of sea-level rise.[79]

Similar to today, at the time of the Aboriginal story about Bremer Bay there was 'a large shallow lake not far from the sea'. The local people were spearing so many fish there that the birds were going hungry, so their leaders – the willie wagtails – implored Marget, a water spirit, to help. Innately indolent, Marget declined to assist, so

> ... *a number of the Willy Wagtails got a long, slender stick, and drove it into one of the mud springs near the lake ... The stick slowly sank in the ground ... and disappeared, but they got another thicker stick and put it on the top, and did the same thing over again. This time the bottom stick touched the sea which ran under the lake and the water gushed up and ran into the lake.*

Stirred into action by the birds' initiative, Marget went to their aid so that 'the sea bubbled and roared through the hole in the lake made by the long sticks'. But the people were not deterred by what had happened, so the willie wagtails again pestered Marget, 'who made the hole bigger ... and the sea roared in harder than ever, making the lake overflow', which caused people to flee the area.[80]

Barrier beaches are occasionally breached, invariably re-forming subsequently, so the existence of a sizeable plug of sediment at the entrance to Wellstead Estuary today is no proper measure of either its longevity or its stability. Such breaches may occur during floods or storms, and certainly do not require a sustained change in ocean level. That said, the details in this narrative about there having once been a lake into which the sea roared are consistent with the sea level rising above a particular threshold and entering for the first time an area of low ground, occupied perhaps by a freshwater marsh that local people valued as a source of sustenance. A sea level around 3m (10ft) lower than today would have rendered such a scenario true.

Almost at the end of our circumnavigation of Australia, we now move east again to one of its most forbidding coasts, that of the **Nullarbor Desert**.[81] The word Nullarbor is not an Aboriginal one, but derives from Latin and means 'no trees', still an accurate characterisation of the appearance of this extraordinary landform that covers a bit over 2,000km^2 (772mi^2), about the size of Kansas, USA, or England and Scotland combined. The limestone geology of the Nullarbor surface tells us that it was originally a part of the ocean floor that, around 14 million years ago, was pushed upwards above the ocean surface. Exposed for the first time to the agents of subaerial weathering and erosion, especially wind, its surface irregularities became smoothed and it began to look as it does today – a vast, flat, treeless plain stretching as far as the eye can see in every direction.[82] Except south.

The southern fringe of the Nullarbor, where it meets the sea in the Great Australian Bight, forms a spectacular set of steep cliffs 50–90m (165–295ft) high, which – in most places – drop to the edge of the ocean itself. There is no coastal flat, just pockets of fallen rocks waiting to be pulverised and removed by the waves that smash relentlessly into the bases of these cliffs, causing them – over many hundreds of years – to slowly recede. The only significant exception to this along the 800km (500 mile) length of the Nullarbor coast is the Roe Plains, a low coastal flat that affords us a glimpse of how the area looked when the sea level was lower during the last ice age.[83]

Aboriginal people have occupied the Nullarbor for tens of millennia, their societies evolving a resilience to the harsh conditions that few others could emulate. The Wati Nyiinyii (or Zebra Finch) people of the Nullarbor have a *tjukurrpa* (oral story) that tells of an old man journeying through the desert, deliberately uprooting every mallee eucalypt tree he comes upon. This was something that threatened the survival of the entire community, for Aboriginal people wanting to drink would often dig up part of the long roots of the tree, and chop them into short lengths before draining the water they contained into bark containers: a practice that would not kill the tree yet would slake their thirst.[84] Uprooting a mallee was reckless, and in the story all the water that was lost as this perverse old man continued on his way drained into the ground, just what happens in limestone country, 'creating a huge flood to the south', where the Nullarbor cliffs lie. The Wati Nyiinyii were obliged to travel en masse to the coast at Eucla to try and stop the encroaching flood:

> The Wati Nyiinyii then pour over the Eucla escarpment, rather like an army of ants ... Once the Wati Nyiinyii reach the sea, they begin bundling thousands of spears to stop the encroaching water. These bundles were stacked very high and managed to contain the water at the base of what is today the Nullarbor (or Bunda) cliffs.[85]

This story may be a recollection of a time when the sea level was lower and a great plain, now covered with ocean, stretched out from the base of the Nullarbor cliffs. As the sea level rose, local groups like the Wati Nyiinyii could not fail to see what was happening: 'individuals thirty years old might have lived through the destruction of a mile [1.6km] of their coastal territory'.[86] Searching for an explanation they sought, as is common, one involving inappropriate life-threatening behaviour that deserved retribution. Yet so concerned were the Wati Nyiinyii at the loss of land that they may have attempted various practical and supplicatory responses, the former including construction of a wooden fence (with spears used as pickets) to try and halt the 'encroaching water' at the cliff foot.

Given the imprecision of the story and the comparatively gentle slope of the sea floor beyond the Nullarbor cliffs, it is difficult to know how far offshore the shoreline might have been at the time when the Wati Nyiinyii first noticed it moving landwards. Were the sea level 10m (33ft) lower, the shoreline would be several kilometres further south here; were the sea level 50m (164ft) lower, the shoreline would lie tens of kilometres off the modern shoreline.

The final story to complete our narrative circumnavigation of Australia comes from **Fowler's Bay**. To Matthew Flinders, sailing east along the south coast of Australia in January 1802, this bay was the first point of shelter in several hundred kilometres. Flinders reported that the 'botanical gentlemen' on board 'landed early on the following morning to examine the productions of the country [yet] found the scantiness of the plants equal to that of the other productions; so that there was no inducement to remain longer'.[87]

Like MacDonnell and other bays along the south coast of the continent, Fowler's Bay is formed by a resistant hard-rock headland protecting an arcuate bay, the hinterland covered with scrub typical of areas fringing Australia's southern deserts. Several versions of a story recalling a time when the coast here became drowned are extant. A detailed one is told by the Bidjandjara people of the Great Victoria Desert.

Malgaru and Jaul were two brothers travelling south from the 'desert' country ... Margaru, the elder, had a kangaroo skin waterbag, as well as two firesticks; but he would not give the other any water. Jaul became thinner and thinner, and his throat more parched. Eventually they came to a place near the south coast – Biranbura, west of Fowler's Bay. There was nothing but dry land there. Malgaru hid his waterbag under some rocks, which were dry at that time, although the sea now breaks over them. There the two brothers quarrelled. Malgaru went out hunting, but as soon as he was out of sight Jaul rushed to the waterbag. In a hurry to get at the water he jabbed at the taut skin with his club, making a hole in it. Water poured out. Malgaru came running back and tried to save the bag, but he could not stem the onrush of water. It spread across the land, drowning them both, and forming what is now the sea.[88]

As with the Nullarbor story, it is impossible to know where the shoreline might have been relative to the modern one when the sea level was first seen to be rising. For the shoreline to be 70–100m (230–330ft) seawards of today's, the sea level would need to be about 10m (33ft) lower. A sea level 50m (165ft) lower would see 1–2km (½–1¼ miles) of sea floor emerge here.

Twenty-one groups of stories recalling the drowning of the Australian coast have been related in this chapter. While showing diversity of form and detail, they all report the same thing – that people witnessed a time when the coastline was much further out to sea than it is today, when what are now offshore islands were part of the mainland, and when now-submerged landscapes were dry and occupied by particular types of animal and vegetation. Had these stories all come from a smaller land mass, where the people relating them had routinely mixed with one another, then one might reasonably suggest that they had perhaps come from a single source – perhaps the memory of an event confined in time and space that had not affected the whole land mass, or indeed, other places beyond its shores. Yet with a country the size of Australia – pretty much the same as the conterminous United States or Europe – no such assumptions are possible. Not only are the stories localised, in the sense that they refer to the geography of the areas in which they were collected, but their original tellers were effectively isolated from each other, and had been for tens of millennia – rather as the native inhabitants of Oregon and Florida, or Scotland and Turkey, were in pre-modern times.

So the most parsimonious interpretation of these 21 groups of stories is that they developed independently of one another, arising from eyewitness observations – accumulated over perhaps several generations – of the rise of postglacial sea level across Australian shores. Given that this rise ceased around Australia some 7,000 years ago, these stories must have endured for at least this period of time, almost wholly as oral traditions passed on with an extraordinarily high degree of replication fidelity from one generation to another – on perhaps more than 350 occasions.[89]

The next chapter provides an account of how the sea level changed in the recent geological past all across our planet's surface. At the end of that chapter, our knowledge of precisely where the sea level stood at particular times in the past is married to each of the 21 groups of stories recounted above, in order to determine – as far as will ever be possible – how long ago the observations that these stories are based on were made. This is an incredible story in its own right.

CHAPTER FOUR

The Changing Ocean Surface

The ocean covers almost three-quarters of our planet's surface and our acquaintance with it varies. Some adults, typically those living in the centres of continents, have never seen the sea and perhaps frame their opinions of it from accounts that describe its extreme conditions – like massive waves washing across low, unprotected coasts and destroying everything and everyone foolish enough to be in their paths. Perhaps such people fear the sea unduly, knowing little of its more constant, largely benign states, which are so familiar to coastal dwellers, who also know well how the ocean surface changes. Indeed, its tidal pulse is so important in deciding the rhythms of coastal livelihoods, from fishing and foraging to trade and tourism, that it sometimes becomes information that no longer needs routine articulation. It is the unheard yet constant heartbeat of long-established coastal communities in every part of the world.

Many of us observe the daily ebb and flow of the tide – a manifestation of a short-lived yet widespread change in the ocean surface – across beaches, up the lower parts of rivers and up the sides of sea walls, without realising that such lunar modulations are in fact just one order of ocean-surface fluctuation. Yet there are innumerable higher order cycles that occur across decades, centuries and millennia. Some cycles can be detected only from millions of years of proxy observation.

Since most sea-level cycles (apart from the daily and monthly) are difficult to identify from casual observation, they remained hidden from science for a long time. Nineteenth-century geologists who believed that they had found evidence for past sea-level changes were often criticised for having mistaken these for land-level (tectonic) movements. It fell to Edmund Suess, who became Professor of Geology at the University of Vienna in 1861, to demonstrate the existence of former cyclical swings of sea level that had affected all parts of the world's coastline. Prompted by his studies of Alpine glaciers, he was able to show that these had periodically advanced in the past, covering much wider areas than they do now, and then receded before once again advancing … and so on. Identifying four such ice ages, Suess correctly inferred that in order to build up extra ice on the land, water needed to be extracted from the oceans; conversely, when ice ages ended, the water from the melting ice was returned to the oceans, raising their levels. Thus colder periods of Earth's history were times of comparatively low sea level, while warmer periods were times when the sea level was higher. Today we live in one of the latter periods – the Holocene Interglacial – a time much warmer (regardless of anthropogenic warming) than most of the time that modern humans have existed, and a time when the ocean surface is tens of metres higher than its average over the past few hundred thousand years. We live, in other words, in a drowned world, one that some of our ancestors 10 millennia ago might barely have recognised.

Modern humans – *Homo sapiens* – first appeared on Earth in the depths of tropical Africa slightly less than 200,000 years ago. They first encountered oceans about 150,000 thousand years ago.[1] Since this book focuses on human observations of long-period changes in sea level, this is an appropriate point from which to commence a discussion of how and why the sea level has changed in the past. Figure 4.1 shows how the sea level has changed within the past 150,000 years, the tail-end of more than 50 climate-driven oscillations (linked to alternating ice ages and intervening interglacial periods), within the past 2.5 million years or so.[2]

What we see on the left of Figure 4.1 is the sea-level rise marking the end of the Penultimate Glaciation (the ice age before the last one) that involved a comparatively rapid rise of sea level from about 140,000 to 130,000 years ago. This is typical of what happens at the end of ice ages – rapid warming causes most of the land-grounded ice to liquefy and pour into

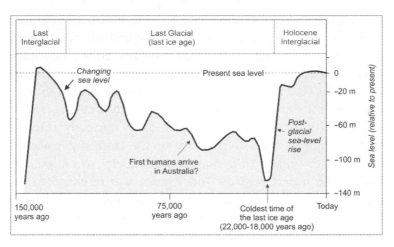

Figure 4.1 *Sea-level change over the past 150,000 years.*

the ocean, causing its level to rise comparatively quickly. The same thing happened at the end of the last ice age, a time that considerably affected modern humans. From about 15,000 years ago until about 7,000 years ago, the sea surface rose rapidly, transforming coastal geographies in every part of the world and forcing people from the margins of the lands. What did these people do? In some places, it is suspected that their attempts to flee inland were resisted by people already occupying hinterlands, so the displaced people quickly taught themselves to build ocean-going vessels and set sail for the ocean horizon, prepared perhaps to find land or perish.[3]

The pattern of sea-level change – and the temperature changes that drove it – shown in Figure 4.1 is typical of most of the climate oscillations of the last few million years, in that cool periods begin comparatively slowly, often with a number of false starts when temperatures (and sea levels) rise for a few thousand years within the overall cooling period. Cooling typically takes far longer, almost 70,000 years in the case of the last ice age, to reach its coldest point, than does the subsequent warming – each of the last two deglaciations took around 10,000 years.

Consider also the amplitude of the sea-level changes shown in the figure. At the coldest time of the Last Glacial, the average global sea level was perhaps 120m (390ft) *lower* than it is today. Consider how that may have transformed any coastline you might be familiar with. Along continental coasts, for instance, the shoreline (where the land meets the sea) would typically have been far further seawards than it is today, maybe by thousands of kilometres. Places that are now islands offshore may then have been part of the mainland, freely accessible by people and animals. Where today coral reefs border shallow continental shelves and island platforms, during the coldest times of the last ice age such reefs were above sea level. They would have formed steep cliffs of limestone perhaps 100m (330ft) high, with only a narrow strip of coastal lowland along their bases. Geography was transformed by sea-level change.

A good example of the profound coastal transformation between the last ice age and today comes from north-west Europe, where the offshore islands now occupied by England, Ireland, Scotland and Wales were contiguous with mainland Europe during the last ice age. Where the English Channel (La Manche) now lies was a broad river valley (shown in Figure 5.1), draining westwards, across the floor of which moved people and any number of species of ice-age fauna, including the woolly mammoth, woolly rhinoceros, a straight-tusked elephant and the spectacularly antlered Irish elk (with an antler span of 3.5m/11½ft), the remains of which periodically show up at Neanderthal sites in the British Isles. One particularly rich cache of woolly rhinoceros and woolly mammoth bones occurs in the ravine of La Cotte de St Brelade on the island of Jersey in the English Channel, having accumulated there at a time when Jersey was merely an area of high ground rising above the surrounding riverine lowlands. One theory has it that herd(s) of mammoths and rhinos, which here date from at least 25,000 years ago, were periodically driven off the cliffs at La Cotte by people as part of a well-planned hunting strategy.[4]

Further north-east, where the North Sea Basin (between eastern England and the Low Countries of north-west Europe) now lies, there was during the last ice age a land mass, named Doggerland today, the form of which has been convincingly reconstructed and the ways of life followed by its inhabitants plausibly imagined.[5] Even before we knew that the sea level in the past had once been lower than it is today, we had an inkling from Doggerland that this must have been so. For centuries, commercial fishermen working the higher parts of Dogger Bank – the now-submerged island that was the last emergent vestige of Doggerland – pulled up animal (including mammoth) bones and human artefacts in their nets, wondering no doubt how these might have reached places on the ocean floor more than 100km (60 miles) from the nearest sizeable land mass.[6] More recently perhaps than you might realise, a major discovery about one of Europe's most ferocious ice-age predators was made here.

On 16 March 2000, the Dutch trawler *UK33* was fishing in the North Sea about halfway between IJmuiden in the Netherlands and Lowestoft in England. In one of the nets the fishermen pulled up was part of an animal jawbone with teeth that appeared sufficiently unusual to merit scientific attention. It turned out to be part of a mandible of a sabre-toothed cat, one of the fiercest felids ever to roam northern hemisphere lands.[7] While the discovery made headlines on account of its rarity, the real repercussions of this find came later, after its age was determined at around 28,000 years ago – almost 300,000 years later than the time it had previously been thought that the last sabre-toothed cat in Europe had died. The implications are that this tenacious predator merely retreated to warmer Mediterranean lands during the ice ages in Europe, then moved north once again as things warmed up, roaming coastal lowlands like Doggerland in search of prey animals sometimes many times its size.

The lower sea levels of the last ice age would have transformed the geography not only of continental margins but also of island worlds. At times of lower sea levels, islands in the world's ocean basins would have been larger and more numerous than they are today. Several convincing models have been developed about faunal migration across the world's oceans, demonstrating how animals used islands as stepping stones, something that would not be viable in today's drowned world. It has even been proposed that ice-age humans crossed the entire Pacific Ocean – the world's largest – from west to east to reach the Americas long before the better studied migrations via the Bering Strait began about 14,000 years ago.

A well-documented example of the effects of postglacial sea-level changes on island peoples comes from the Channel Islands off the coast of California, USA. They were first occupied by people who reached them by crossing the 7km (4½ mile) wide Santa Barbara Channel (which was narrower than it is today) about 13,000 years ago, when the postglacial rise of sea level in the area was well underway. While today there are four main islands in the group (San Miguel, Santa Rosa, Santa Cruz and Anacapa), at this time of lower sea level these were all one island – posthumously named Santarosae – that also incorporated a sizeable area of exposed insular shelf. Surrounded by pristine reefs, brimful with food, the sight of such an island offshore must have stimulated bolder inhabitants of the increasingly crowded mainland coast to take to their boats. Some evidently succeeded in making the crossing, but their descendants were ultimately disadvantaged by the move, for as sea levels continued rising so Santarosae became smaller and increasingly isolated from the mainland, eventually breaking up into the four constituent islands we see there today, reducing the livelihood options available for their inhabitants.

The long oscillatory fall of the sea level from the end of the Last Interglacial to the Last Glacial Maximum (see below) transformed the Earth's climates and its coastal geographies. Yet it was the cooling that drove this sea-level fall that had more widespread effects, bringing lower temperatures to places where plants and animals were accustomed to warmer conditions. A glimpse into this cooling world is provided by the hypersaline Dead Sea, the surface of which lies more than 400m (1,310ft) *below* the ocean surface. Bounded on the east by the Jordan Plateau and to the west by the Judean Mountains, the Dead Sea is a terminus for water entering along the Jordan River. For this reason, the sediments accumulated on the floor of the Dead Sea over tens of thousands of years have the potential to tell us a lot about the changes in climate that have affected the region in the past.

We know that during the cooler periods of the past few million years – the ice ages – the region became much wetter, causing the Dead Sea to expand to accommodate double today's rainfall. Conversely, during interglacial periods such as the one we live in today, the region was drier and the Dead Sea contracted. Examination of the lake-floor sediments has allowed insights into the shorter term changes in the lake level driven by precipitation changes across the region. Thus we learn from recent research that around 87,000–93,000 years ago there was a rapid drop in the lake level probably driven by the abrupt onset of some 6,000 years of drier, warmer conditions. Because this warm interval occurred within the overall cooling marking the early part of the last ice age, it is referred to as a stadial. The Dead Sea came close to drying up completely. Then the ice age resumed in earnest, the period between 75,000 and 87,000 years ago being one of conspicuously high lake levels, corresponding to a time of cooler conditions marked by ice-sheet build-up across much of the northern hemisphere. This period of cooler conditions within the overall cooling is known as an interstadial. Long interstadials punctuated by shorter stadials are characteristic of the long, slow cooling that marked Earth's transition from the Last Interglacial to the Last Glacial Maximum.

At last, around 22,000 years ago, 70,000 years of climatic prevarication came to an end and the world was plunged into the coldest time of the last ice age – a time known as the Last Glacial Maximum (LGM). Lasting for perhaps only a few millennia, depending on where you were in the world, the LGM would inevitably have posed challenges to the way our ancestors had been living up to that point. Take those in Eurasia, from Ireland to Kamchatka, a region for which there are excellent data about the links between LGM climates and vegetation. The dominance of temperate forests in the region came to a comparatively swift end at the start of the LGM. Cold-adapted forests and tundra dominated during the LGM but were gradually pushed northwards after its end, being replaced in lower latitude areas by temperate forests.

The LGM was not only the coldest time within the past 150,000 years, but also one when the sea level was lowest – at an average of around 120m (400ft) below today's levels. What this meant was that land masses were bigger than they are today, often with land connections where now there is ocean. This world was generally easier for terrestrial plants and animals to get around in, although due to the cold conditions many of those we are most familiar with today were often confined to refugia, places where environmental conditions were fortuitously configured to allow a particular species – or group of species – to survive.

Consider the thinhorn or dall sheep (*Ovis dalli*), a wild sheep inhabiting cooler parts of North America to which it is native. Study of thinhorn sheep genes shows that this species survived through the LGM in this region by occupying ice-free refugia. Research suggests that different subspecies of the sheep evolved at this time within different refugia, and that when the ice was gone they reunited and interbred.

It was not only ice that pushed animal and plant species into refugia, but also the lower sea level that transformed many coastlines, so that particular habitats temporarily disappeared in certain places, becoming re-established only after the end of the LGM when the sea level began rising. A good example is provided by tidal estuarine habitats along the coasts of California and Baja California (Mexico). Before the LGM (as is the case today) these habitats were occupied by several species of estuarine fish, but during the LGM, when the sea level in this part of the world dropped to 130m (430ft) below its present level, these habitats vanished – and the fish along with them. Similarly to the thinhorn sheep, the fish species survived in two refugia north and south of the area, 1,000km (625 miles) apart, coming back together only after the sea level rose and tidal estuarine habitats were restored to coastal California.

While we think of the LGM as the coldest time of the last ice age, it is also important to appreciate that cold did not affect every part of the Earth's surface. While polar and temperate climate zones may have expanded towards the Equator, the effect of this was to reduce the size of tropical regions – concertinaing them, if you like – but they were still comparatively warm and hosted most tropical species of plant and animal that had been able to move there from higher latitudes. Yet in many mid-latitude regions (not Australia), the last ice age was marked by wetter climates that in many such places opened up new opportunities for living things, including humans.[8]

As the LGM drew to its close and Earth's surface temperatures began to rise, living things may have drawn a collective sigh of relief that the millennia of hard times were finally at an end … and that a warming world would be more suffused with promise and opportunity. But this was not to be, for living things (as a rule) deplore environmental change, especially when it happens quickly. This was to be inevitable in the postglacial world, in which temperatures and sea levels not only rose comparatively rapidly, but also did so in bursts punctuated by times of stasis or even short-lived cooling and sea-level falls. It was not to be an easy road ahead for many species, and some of the conspicuous megafaunal extinctions (see Chapter 6) that occurred during the half-dozen millennia following the end of the LGM have been blamed on rapid changes of climate.[9]

The idea that rising temperatures in the aftermath of the LGM would have melted most of the massive continental ice sheets, with the meltwater flowing down rivers into the oceans, the surfaces of which would rise in response, is fine for conceptual and illustrative purposes – but not ultimately realistic. There are two main difficulties.

Firstly, when a thick ice sheet develops on a continent, the weight of the ice actually causes the continent to sink; conversely, when the ice melts, the continent rises – a phenomenon known as isostatic rebound. So the coastlines of continents from which ice sheets were disappearing during deglacial times were actually rising (as was the sea's surface) at the time, making it very challenging to isolate the precise magnitude of actual sea-level rise in such places. For this reason, oceanic islands – generally far from formerly ice-covered continents and with only narrow insular shelves – have for decades been favoured by scientists interested in measuring the exact magnitudes of postglacial sea-level changes.[10]

Secondly, water from melting ice sheets on the land did not always easily reach the ocean. Sometimes natural dams formed in the narrow, deep valleys around the fringes of the former ice sheet, causing meltwater to pond behind them – much as today's intentionally constructed dams do. In other situations, where ice sheets had created massive depressions in the centres of continents, the melted ice could not easily escape to the ocean. In both such situations, enormous meltwater lakes formed in the centres of many formerly ice-covered continents, enduring for sometimes thousands of years before typically breaching a dam or incising a bedrock barrier, which led to their rapid emptying. The meltwater floods poured huge volumes of fresh water into the oceans, often causing abrupt ocean-ecosystem changes and even circulation changes, besides sometimes raising ocean levels quite quickly over a short period of time. Such floods are also implicated in short-lived climate changes that affected the entire planet, posing huge challenges for living things while momentarily, it seems, interrupting the progress of postglacial sea-level rise and the warming that ultimately drove it.

Dome C in Antarctica is a forbidding place, one of the coldest areas on Earth, with summer temperatures rarely warming above -25°C (-13°F) and winter temperatures often plummeting to below -80°C (-112°F). It hardly ever rains, and we are the only sapient creatures who elect to live. Scientists' interest in Dome C comes from the fact that it is more than 3,200m (10,500ft) above sea level, one of the highest parts of this massive ice-covered continent, which means that the ice here is thicker than almost anywhere else. In addition, because ice accumulates regularly albeit very slowly in such places, Dome C sits atop one of the lengthiest and most complete archives of the climatic history of Antarctica, each layer of accumulated ice containing information about the climate of the time it was laid down. And because Dome C is so close to one end of the Earth – just 1,670km (1,038 miles) from the South Pole – and because Antarctica is surrounded by ocean, its climate registers the effects of climate changes across the entire planet. Indeed, the record from ice cores through the ice below Dome C has given us solid information about the effects of meltwater bursts on our planet's past climate.

In 1996, a 3,270m (10,728ft) ice core was extracted from Dome C by members of EPICA, the European Project for Ice Coring in Antarctica. Analyses of the core allowed insights into the ways in which the Antarctic climate has changed over the last 800,000 years. One of the most remarkable findings was that of temperature, proxied by the hydrogen isotope deuterium. It could be shown that particular episodes of higher temperature here coincided with those of lower temperature in the northern hemisphere. We are not talking about the major swings of temperature that distinguish ice ages from warm interglacial periods, but about shorter duration changes of perhaps a few millennia at most. The unexpected finding that changes in the climates of the northern and southern hemispheres were out of phase led ultimately to a better understanding of the effects of the catastrophic emptying of meltwater lakes.

One of the earliest periods of rapid temperature change to be recognised during deglacial times, when the Earth was transitioning from ice-age conditions to interglacial ones, is called the Younger Dryas and lasted for perhaps 1,000 years in most parts of the world around 12,000 years ago. In much of the northern hemisphere, the start of the Younger Dryas saw an abrupt return to full-glacial conditions. Conversely, when it ended here, temperatures rose abruptly – almost 10°C (18°F) in 10 years in Greenland – after which the slower warming trend characteristic of the postglacial period resumed. The EPICA core showed something quite different, namely that within the same period of time in Antartica temperatures rose, not fell. Why was this?

The most plausible explanation for it to date involves the thermohaline circulation of the world ocean on which we depend today (as in the past) for moderating extreme climate conditions in many places. The thermohaline circulation – or oceanic conveyor belt – is driven by variations in seawater density, controlled largely by ocean-water temperature (the thermo part of the name) and salinity (the haline bit), and has effects on every part of the ocean. Much of the thermohaline circulation occurs deep within the world's oceans, but it comes to the surface in the North Atlantic. Here warm salty water is transported from the tropics north to polar waters, where it loses heat and sinks. Now imagine a vast meltwater lake in the centre of Canada, one of a number of outsized puddles left behind after the melting of the Laurentide ice sheet, which covered this part of North America during the LGM. The lake in question formed along the margins of the shrinking ice and may have covered an area of $440,000km^2$ ($170,000mi^2$) – larger than any lake in existence today – about 13,000 years ago. As the ice-sheet margins continued melting, so a passage to the sea finally opened for this brimful lake. The resulting outburst flood saw so much fresh water enter the North Atlantic (via the Mackenzie River in the Canadian Arctic Coastal Plain) that it temporarily shut down this key part of the thermohaline circulation, soon bringing the rest of the great oceanic conveyor belt to a grinding halt. This may have been the ultimate cause of the Younger Dryas event.

The result of the shutdown of the thermohaline circulation would have been abrupt cooling of the northern hemisphere and coeval warming of the south, exactly what polar ice-core data from places like Dome C suggest. The effects of such abrupt climate change on human societies were many and varied, but few were spared its effects. There is a school of thought that regards the Younger Dryas cooling as having forced people into experimenting with agriculture and animal domestication – the onset of cold conditions gave them little choice were they to survive. Necessity is indeed the mother of invention. In support of this suggestion, it has been noted that the earliest known ages for nascent food production in many places do appear to coincide with the Younger Dryas event.

The effects of the Younger Dryas cooling on the sea level are generally more difficult to detect. In most parts of the world's oceans, it seems that this rapid cooling temporarily slowed the rate at which the postglacial sea level had been rising. When the Younger Dryas ended, the rate accelerated back to what it had been before. Yet the Younger Dryas was preceded – and the majority scientific view at the moment appears to regard the juxtaposition as fortuitous rather than causal – by a rapid global rise in sea level, known as Meltwater Pulse 1A (MWP-1A). Occurring some 14,500 years ago, MWP-1A involved a sea-level rise of some 15m (50ft) in a mere 340 years.

There are different ideas about the cause(s) of MWP-1A, with some evidence pointing to Antarctica as the source of the meltwater, and some to North America. One of the most compelling ideas is that the low ice 'saddle' between the two major North American ice sheets of the time – the massive Laurentide in the centre and east, and the smaller Cordilleran in the west – may have collapsed abruptly, tipping massive amounts of ice-choked water into both the North Atlantic and the Pacific Oceans, quickly reducing their surface temperatures and raising their surfaces.[11]

More recently, about 8,200 years ago, the last of the great North American meltwater lakes – Lake Agassiz-Ojibway – smashed through its ice dam in Hudson Bay, generating another massive injection of fresh water to the North Atlantic and shutting down the thermohaline circulation once again – this time for some 400 years. Temperatures on land masses around the North Atlantic plunged; central Greenland dropped by as much as 8°C (14.5°F) and Western Europe by perhaps 3°C (5.5°F) within this period.[12] The evidence for what happened during this 8,200-Year Event is clearer than it is for the Younger Dryas, even to the extent that it has proved possible to directly measure the effects of the outburst flood on the sea level. For example, along the northern coast of the Gulf of Mexico, where the Mississippi Delta lies, the sea level rose perhaps 2.2m (7¼ft) within the 130-year duration of the 8,200-Year Event, but the immediate response was a near-instantaneous 56cm (22in) rise.[13]

Ultimately no coasts in the world were immune from the effects of these rapid meltwater-fuelled bursts of rising postglacial sea level, but the record is essentially untraceable in places furthest from where they occurred. The author knows of no clear records of such rapid sea-level rise events from Australian shores, for instance, although faint hints have been found in coastal sediments dating from that time in places like Singapore, the Sunda Shelf (where island South-east Asia now lies) and Hong Kong.

The point of this discussion is to show that postglacial sea-level rise was not a uniform process, unvarying in rate, but one that was punctuated by bursts of sudden rise, which posed the greatest challenges to living things, including our coastal-dwelling ancestors. While we are on the subject, there is a lesson for the future buried in what might appear fairly obscure musings about the distant past. For if the climate were to change to such a degree that large terrestrial ice masses – like those in Greenland and Antarctica – became unstable, slipping into the world's ocean, this could indeed cause similarly calamitous changes along many of the world's coasts.[14] History warns us what might happen, but there is no purpose in losing sleep over it.

Whether the sea rose in swift bursts or more slowly and uniformly, the effects of postglacial sea-level rise on the world's coastlines varied depending on two things.

The first of these relates to whether or not the land around which the water rose was itself moving. This makes a big difference, since should a coast already be sinking, sea-level rise will have a multiplier effect. Conversely, should an area of coastal land be rising, this will reduce – it may even reverse – the effects of sea-level rise. During postglacial times, some coasts – relieved of their thick, icy overburden – rebounded upwards, often rising faster than the sea level was rising. For people living along the coasts of Norway and Sweden, for instance, unlike those in most other parts of the world at the time, this meant that the land was emerging relative to the ocean surface – even though both were rising. As an example of how this affected coastal people in these areas, take the Viking settlement at Borre in Norway, a place acknowledged in the Norse Sagas as a royal burial site, and of such great renown among the Norse seafarers that it might come as no surprise to learn that an entire ship was buried there. Active largely in AD 600–1000, the Borre settlement now lies 5m (16ft) above the coast, raised there by isostatic rebound. It is a fossil world that is harbourless and therefore now useless for its original purposes, with the remains of wharves and jetties forming minor features in an inland landscape. Many Scandinavian coasts are still rising, still rebounding, and this action will continue to affect their inhabitants and the ways their descendants live.

Land can sink as well as rise. Most oceanic islands are sinking, but typically so slowly that they hardly affect the observed rate at which the sea level periodically rises. A contemporary example, from a world in which there has been an upward trend of sea level for more than a century, may make more sense. Many of the world's longest established coastal or delta cities have a long tradition of drawing fresh water from underground aquifers to quench their inhabitants' thirst and that of their industry. Over time, extraction of groundwater causes the spaces between the particles in the delta sediments below these cities to collapse, as the water that formerly kept them buoyant is withdrawn. Even when underground aquifers have hardly been tapped for water, cities built on deltas – little more than massive piles of unconsolidated sand and gravel – also tend to sink because the sheer weight of the delta (and the city on top of it) can cause both the underlying solid crust of the Earth to deform and the loose sediments to become compacted. Collapse and compaction at a giant scale lead to gradual ground-surface sinking, a phenomenon that affects cities like Bangkok and Shanghai as much as New Orleans and Venice. For example, between 1956 and 1965, parts of Shanghai subsided at an average rate of 83mm (3¼in) each year, a rate that has somewhat lessened because of aquifer recharge but is still enough to amplify the effects of flooding in the downtown area of the city, where subsidence rates remain highest.

The second cause of variation in the ways in which postglacial sea-level rises affected the world's coasts is of course their character, particularly whether they rise gently, perhaps almost imperceptibly, from beneath the ocean surface, or far more boldly. Unfortunately for most people (like the author) who live on the coast, the majority favour coasts that are gently sloping – well designed, it seems, for accessing other places, for growing crops and for building cities. And why ever not? Who wants to build a settlement and roads that cling to a rocky cliffy shore, let alone try and eke out a livelihood there, when other, more salubrious options are on offer? Yet coastal plains, which are commonly low and extensive, are the most vulnerable of all coastal types to sea-level rise. Over the tens of thousands of years across which we can reconstruct the history of coastal change, it is such places that come and go most frequently, often appearing with a flourish of civilisation, then disappearing with its collapse, while coasts that are high and where there is little lowland may have cultures that endure far longer.

The sea-level changes to which people occupying almost every part of the world's coastline (except perhaps the Mediterranean and some parts of the North Atlantic) have become accustomed over the past few millennia are quite different to those that characterised the postglacial period, when melting land ice drove sea-level rise. For land-ice melt attributable to rising Earth-surface temperatures linked to Earth's orbital changes – the pacemaker of the ice ages – is considered by most sea-level scientists to have stopped about six millennia ago. Yet since then the sea surface has continued to change. It has not done so by very much compared to its period of postglacial rise, yet sufficiently at times to have caused noticeable changes to coastal environments – often forcing changes to the ways in which their inhabitants live.

It is generally considered that sea-level changes within the past 6,000 years have been temperature driven – what are termed steric changes. Within this period, temperature has been the main driver of sea-level change over decades and centuries; when temperatures fall, the sea level falls; when temperatures rise, the sea level will also rise. This close relationship between changes in temperature and sea levels is explainable by two processes. First, when you heat water it expands slightly. Thus, when the upper parts of the ocean are warmed over many years, they expand and the ocean surface rises. Conversely, when ocean-surface waters cool, they contract, occupying less space, so the sea level falls. Secondly, when temperatures rise, there is a net loss of land ice, with the meltwater produced ending up in the ocean and causing the sea level to rise. Conversely, prolonged temperature fall will eventually cause land ice to increase in volume, lowering the ocean surface. The emphasis on land-based ice in both scenarios is important because sea (floating) ice cannot alter the sea level whether it forms or melts.

Since this period – the last 6,000 years or so – is somewhat marginal to the main focus of this chapter, a couple of examples suffice to demonstrate the effects of comparatively small sea-level perturbations on human affairs, which have become more and more sensitive to such external pressures as we approach the present day.

Superbly evocative examples come from some of the world's larger river deltas, places where the history of *relative* sea-level changes – those reflecting both ocean-water volume changes and land-level movements – can be remarkably complex. Consider the Yangtze (Changjiang) Delta in China, which was first occupied by people sustained by rice agriculture at least 7,000 years ago. Sea-level oscillations at the end of the postglacial transgression alternately drove people off these lands when the sea level was higher, and allowed them to reclaim their former territories when it fell subsequently. In the latter case, the new settlers favoured wet-rice cultivation in this region, which may have been the place from which rice agriculture then dispersed throughout East Asia.[15] The graph below (Figure 4.2) shows neatly how sea-level changes in the Yangtze Delta can be apparently linked to the rise and fall of civilisations there several millennia ago.

A similar situation once obtained in northern Europe, where alternations in the occupation of the Vistula River Delta (northern Poland), which empties into the Gulf of Gdańsk, within the last 5,000 years were closely linked to sea-level changes. The rising sea level at the start of this period created coastal wetlands, a time followed by several hundred years of comparatively stable sea level that allowed people to develop an entire system of livelihoods based on offshore

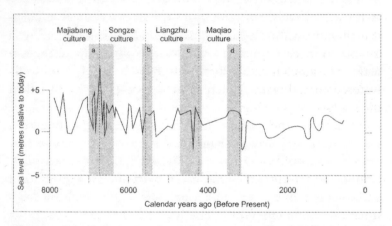

Figure 4.2 *Relative sea-level changes and cultural declines (marked by shaded bands) in the Yangtze Delta, China. Note that these sea-level changes are termed relative as they include both ocean-water volume changes and changes in land level, here typically linked to alternate periods of deltaic build-up and sinking. The collapse of the Majiabang culture came about when the high sea level led to groundwater flooding; that of the Songze culture also because of flooding associated with the high sea level, but in addition the intense variability of the ocean surface; that of the Liangzhu and Maqiao because of water-table rise and the expansion of the Taihu Lakes, which caused widespread inundation.*

fishing and hunting in deltaic forests of alder, hazel and elm. As the sea level rose again subsequently, so the delta shoreline moved landward and the people's livelihoods became dominated by fishing and sea hunting; it seems to have been just too wet underfoot for most of the year to support similar forests. Agriculture took a hold in the area much later as a gravel barrier, which shielded much of the delta from the ocean, allowed the return of forests that were later cut down to open up land for farming.[16]

Another example of how sea-level change caused widespread and enduring changes to human societies comes from the tropical Pacific islands. Around 700 years ago – AD 1250–1350 is the best estimate – there is evidence that the sea level *fell* slightly across this vast region. It may have only fallen by a few dozen centimetres, but that was enough, it appears, to cause massive problems for the islanders who had up until this time lived almost entirely along island coasts, eating well for the most part from the rich sources of readily accessible seafoods, supplemented by foods grown along the well-watered coastal plains. Then came disaster. The slight fall in the sea level was enough to expose the surface of the offshore coral reefs – the central link in the seafood-production system – and to lower on-land water tables sufficiently so that crop roots could no longer easily reach them. Reefs died, lagoons went unflushed by seawater, crops failed – and within a few generations people had much less food to eat. In such a situation, when everything else has failed, rather than starve you naturally turn your gaze towards your neighbours. Maybe to eat ... but more likely to get your hands on the foods they have to eat, something they would be unlikely to give up without a fight.

Thus the upshot of this sea-level driven food crisis was the outbreak of conflict throughout the tropical Pacific islands. By about AD 1400, people living on high islands across this third of the Earth's surface had all but abandoned coastal settlements in favour of new ones established in fortifiable locations, usually inland and upslope.[17] In island nations like Fiji, where much of my research into hillforts – or *koronivalu* (literally towns-of-war) as Fijians call them – has been concentrated in recent years, it is clear that conflict following this food crisis endured for hundreds of years.[18] When the British began settling in Fiji in earnest around the 1850s, some Fijians were still occupying hillforts. Consider this 1870 report by an Australian newspaperman about what he saw along the coast of Viti Levu Bay, in the north-east of Fiji's largest island:

> *In the distance [are] the houses of the mountaineers, perched curiously on the apex of rocky pinnacles, a position singularly secure from invasion... Walking around the base of the bay in a northeasterly direction we came to a thickly populated town on the crest of a hill, east of which could be seen another native village, built on an immense rock. The natives who were all clustered together on a little plateau in front of the town received us very nervously...[19]*

The plausible connection between climate-induced sea-level fall and societal upheaval for Fiji and other Pacific Island groups about 700 years ago provides an excellent example of how coastal dwellers – in whatever part of the world they may live – pay a price for the advantages of their location, a cost that is occasionally charged when the sea level changes.

The ocean surface has been rising, on and off, for the better part of the last two centuries. The upward trend is clear, although doubters commonly exaggerate the significance of short-lived periods of stasis or even fall. These are to be expected in any natural system being closely monitored over long time periods. The global sea level is currently rising at an average rate of 3.2mm each year, somewhat higher than its average rate of rise of 1.7mm per year over the past 100 years or so.[20] Observations explain 87 per cent of the 54mm (about 2in) of sea-level rise between 1993 and 2010 as being caused by the two processes mentioned earlier. Thermal expansion was responsible for 37 per cent, meltwater from glaciers and ice sheets contributed 50 per cent, and the remaining 13 per

Above: Eruption of the Kavachi underwater volcano, south of Gatokae Island, Solomon Islands, January 2005.

Left: A hand-coloured lithograph, dated 1886, of an eruption near Tonga, almost certainly that of Fonuafoʻou (Falcon Island), by an unknown artist. The resemblance of the eruptive clouds to giant fish is likely to have inspired Pacific Islander stories about islands being fished up from the sea floor, often by the mischievous demigod Maui.

Above: The March 1847 eruption of Mt St Helens, in the Cascade Range of Washington State, US, painted by Paul Kane. This image shows details of the eruption cloud that was interpreted by native Americans in the area as that of a giant eagle pecking at some prey on the mountain's flank. The yellow beak and red head of the eagle can be seen poking out from its mass of dark feathers.

Left: *La Fuite du Roi Gradlon* (The Flight of King Gradlon) painted by Évariste-Vital Luminais, 1884. King Gradlon's daughter (Dahut) is possessed by a demon. Stealing the keys to the sluice gates, she let in the ocean and flooded her home city of Ys. It is possible that this legend is a distant memory of a time when a city off the coast of France became submerged.

Above: Remains of a submerged forest exposed by storm-wave erosion in February 2014 between Borth and Ynyslas off the modern coast of Wales. When the sea level was lower, as it was several thousand years ago in this area, lowland forests such as this one thrived.

Left and below: Eroding cliffs, Dunwich, UK, showing the remains of the 13th-century All Saints Church teetering on the cliff edge. The church remains fell into the sea in 1922.

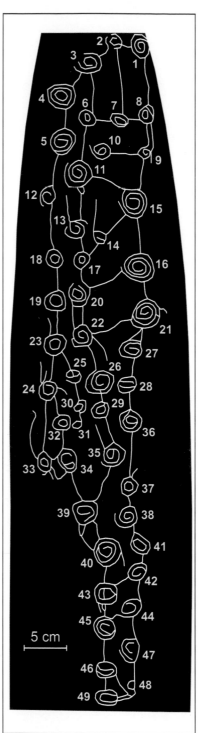

1. Labbi-labbi; 2. Taŋanja;
3. Liuwiriŋa; 4. Kunnamannera;
5. Maiyada-maiyada; 6. Wirra-wirra;
7. Kirindji; 8. Kanandibarro;
9. Markodarindja; 10. Kampanbarro;
11. Wirrkaldjarra; 12. Pinna;
13. Luwano; 14. Kira;
15. Tjul'tjun'waridji; 16. Ŋandju;
17. Tildi; 18. Wakilbi; 19. Kuna;
20. Pintinba; 21. Yinindi;
22. Yalbirrimanno; 23. Tanda;
24. Kurandal; 25. Palta; 26. Kura;
27. Biŋbiyaŋ; 28. Tjipallalla;
29. Yirabanda; 30. Ŋaŋgalli;
31. Yappadarra; 32. Timbabiddi;
33. Yuldumallo; 34. Kunagarri;
35. Mukubanda; 36. Mari-mari;
37. Karruwildji; 38. Wallabarrarba;
39. Kiribarro; 40. Yanno;
41. Wangadjarro; 42. Wornba;
43. Tjimarri; 44. Kunaŋanno;
45. Wirrarigulong; 46. Ŋanneriyoŋo;
47. Miltji-miltji; 48. Papulba;
49. Lola.

5 cm

Left: Interpreted detail of a carved spear-thrower (*llanguro*) of the Pintupi (Bindibu) people of the central desert of Australia. Representing the pathway of the great Snake (*liro*) through the landscape, the design is actually a map showing the locations of 49 water sources, named on the right. Information was given by Tjappanoŋgo of the Pintupi to Donald Thomson in 1957.

Above: The Roman market at Pozzuoli (Macellum di Pozzuoli) in southern Italy was built around AD 200 and has alternately sunk and emerged since then. This is like a barometer of earth-surface movement in the Phlegrean Fields, where the underground movements of superheated water and magma cause periodic localised crustal swelling and collapse.

Left: *Mudra* (or seal) thought to have belonged to a resident of the ancient city of Dwaraka, India. The design is of a three-headed animal; such seals were carried by citizens to identify themselves in times of unrest.

Below: Carved bas-relief at Prambanan Temple, Indonesia, shows the monkey army of Lord Rama (left) carrying stones to make the causeway between India and Sri Lanka. Sea creatures – a giant fish and the legendary *makara* – help construction.

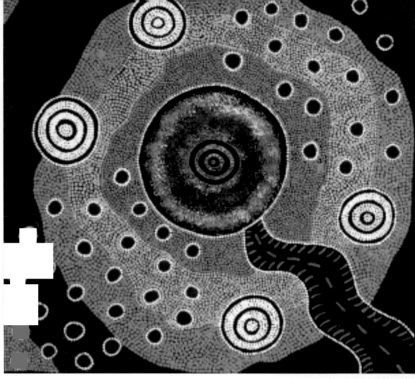

Above: *The Fly Dreaming* (*Ngurriny*) by Australian Aboriginal artist Jane Gordon of Billiluna is a painting that recalls the time when a 'ball of fire' came down from the sky and 'shook the ground', releasing flies from the hole it made and allowing snakes to make their homes in the crater.

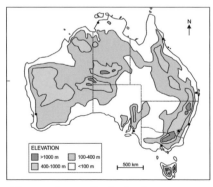

Above: Australia is the lowest continent with very little land more than 1,000m above sea level – and lots well below 100m.

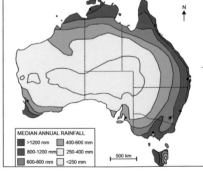

Above: The country is also uncommonly dry with vast deserts in its interior and most rainfall concentrated along its coastal fringes.

Above: The lake below the summit of Tangkuban Perahu, Indonesia. It perhaps formed when landslides dammed the Citarum River almost 16,000 years ago. A similar lake is referred to in local stories as having been created by a man named Sangkuriang to provide a place for him and his prospective bride (actually his mother) on which to sail.

Left: The ancient ovens belonging to the giant Craitbul. This Australian Aboriginal story of the origin of the volcanic craters that today form Mt Schank in Australia appear to corroborate the young ages for its latest eruptions.

Left: The largest meteorite impact crater in Australia, Wolfe Creek Crater, formed during a single event some 300,000 years ago, long before humans arrived in Australia. Yet there are innumerable Aboriginal stories that recall the formation of the Wolfe Creek Crater (or Kandimalal).

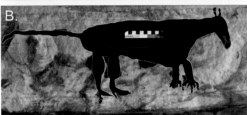

Above: Australian Aboriginal stories of bunyips and other creatures may be distant recollections of extinct megafauna. A and B show a likely depiction of the 'marsupial tapir', *Palorchestes azael*, in the Kimberley region of Australia (scale bar is 10cm long).

Above: Reconstruction of *Palorchestes azael* by Peter Schouten. Palorchestes was a herbivore and probably too slow to escape agile and inventive predators like humans.

Above: Possible representations of *Genyornis newtoni* from the Nawarla Gabarnmang rockshelter in Arnhem Land, Australia. Note the big beak and large claws that were characteristic of this bird. Inset shows a reconstruction of *Genyornis newtoni* by Peter Trusler, used on a 55c Australian stamp.

cent is attributed to the extraction of groundwater by people and its eventual arrival in the ocean.[21]

During the coldest time of the last ice age (the LGM), the continent of Australia was 30 per cent larger than it is today.[22] Where today's shallow submerged continental shelf is widest, the coastline during the LGM was thousands of kilometres seaward of its present position, all land once occupied by people that has since been drowned. Most of what are now smaller offshore islands, including Tasmania, were connected to the mainland. And significantly, there was a continuous land bridge connecting Australia to what is now the large high island of New Guinea, a land bridge where the Torres Strait and the scatter of islands within it now lie. Across this land bridge, people moved freely for most of the time they occupied Australia. They clustered around large fresh Lake Carpentaria into which the ocean, forced by the rising sea level, finally spilled about 10,500 years ago, leading to the development of the modern Gulf of Carpentaria.

Given that people arrived in Australia around 65,000 years ago, tens of thousands of years before the culmination of the last great ice age about 20,000 years ago, we know for certain that they witnessed the subsequent rise of sea level that trimmed back the extremities of their land so swiftly and by such extraordinary amounts. It has been estimated that people living 10 millennia ago on the low-lying coastal plain south of the Nullarbor Desert would have witnessed the shoreline moving landwards at a rate of a metre (3¼ft) each week. And off the northern shores of modern Australia, it has been calculated that during more rapid periods of postglacial sea-level rise, people would have seen the shoreline retreat landward by 5km (3 miles) every year – a startling thing to contemplate. It would be interesting to know to what degree such dramatic rates of change created a millennialist mindset among contemporary Australians. Did they suppose that the sea level might continue rising independently until it ultimately engulfed all of Australia, condemning its inhabitants to a fate like that of people in the time of

Gilgamesh and Noah? Did they accept what was happening as an expression of divine power, an empyrean whim they could not contemplate challenging? Or did they take a more pragmatic view, seeking ways to stop the rise of the waters and halt the contraction of their lands? Much evidence at the moment points to the latter – ancient stories from many parts of Aboriginal Australia recall that people at the time constructed sea defences – and it is likely that such actions were complemented by spiritual activities intended to rationalise and counter what was happening.

We saw in Chapter 3 that there are innumerable extant stories that plausibly recall the nature and effects of postglacial sea-level rise in 21 areas around the fringes of Australia. While extraordinary to ponder, the existence of these stories renders plausible the idea that postglacial sea-level rise had a major impact on the Aboriginal psyche. It is likely that the words of the stories that have come down to us today were merely part of a panoply of beliefs captured at the time in various media – including prose, song, dance and art – which enshrined the memories of sea-level rise and their various explanations in Aboriginal traditions. If the longevity of a particular oral tradition can be taken as a measure of its impact on those who originally conceived it – the eyewitnesses – and a measure of the concomitant need to have it communicated effectively to future generations, then clearly the impact of sea-level rise on the coastal-dwelling Aboriginal peoples who witnessed it was very powerful. But just how powerful? How many years have these stories endured?

To answer this question, we can take as a starting point the fact that the ocean surface – the sea level – reached its present level around Australia about 7,000 years ago. Since that time, steric effects have caused minor oscillations of sea level, but it has not fallen more than 1m (3¼ft) below its present level, nor in most places has it risen much more than 2m (6½ft) above it. So hardly any of the stories recounted in Chapter 3, except maybe those from Moreton and North Stradbroke Islands and that from Bremer Bay, could possibly be based on observations made within the past 7,000 years. Put another way, most of these stories must have endured – as oral traditions – for seven millennia or more.

While we are contemplating the extraordinary proposition that a story – let alone a number of stories – could survive in recognisable form for more than 7,000 years, consider that the comparative stability of the ocean surface within this period would not have allowed any reinforcement of the story using contemporary observations. And yet, it must have appeared increasingly implausible to each successive generation living through these thousands of years that the sea level was ever much lower than it was – as the stories said. Maybe in some communities the stories of the elders were challenged, and the more implausible – like those about the sea level having once risen and drowned lands that are now underwater – were sidelined, and intentionally not transmitted to the next generation. Yet, such speculation aside, it seems indisputable that some groups – perhaps more than 20 of them – were able to effectively transmit drowning stories across a vast slice of time, making their people the greatest of all the oral chroniclers of human history.

As for most parts of the world, the history of postglacial sea-level rise around Australia is known fairly precisely today. A compilation[23] is shown in Figure 4.3 and links time (in years ago) to ocean depths. The exact level of the Australian ocean relative to today for any one year in the past is impossible to determine, *both* because there were apparently minor variations between the levels of the ocean in different parts of the Australian coast at any one time, *and* because the techniques used to determine past sea levels are unavoidably imprecise. For these reasons, in Figure 4.3 the sea level in any one year can be given only as a range. Alternatively, the time at which the sea level reached a particular depth (or height) relative to today is also expressed as a range. Thus, as you can see, the sea level was 60m (200ft) lower than it is today between about 12,300 and 13,100 years ago – an 800-year range is as good as it gets here. Things are a bit better for the time when the sea level was 30m (100ft) lower, something that occurred around every part of Australia within the 500-year period between

9,950 and 10,450 years ago. Finally, as you can also see in Figure 4.3, the rising ocean surface reached its present level (zero metres) some time between 6,350 and 7,700 years ago. We know that there was a lot of spatial variation in this event, something that reflects the difficulty of determining precise

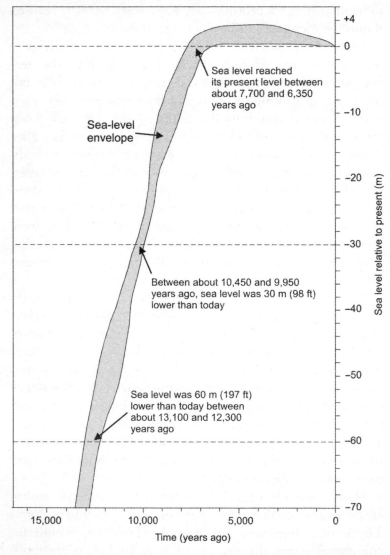

Figure 4.3 *Sea-level envelope for Australia over the last 13,000 years.*

data from the evidence, which (as discussed below) generally becomes more complex about this time.

You might be wondering how anyone can know how much lower the ocean surface was, say, 10,000 years ago. Along almost every part of the world's coastline the answer lies underwater, and while finding it is certainly not as difficult as it was a century ago, when scientists first realised it must be there, it is still a challenge to do so.

Every time it rains, soil and other loose particles – sediments – are washed off the land and into the sea. Sometimes if the particles are fine, they float for a while, but most end up sinking to the ocean floor where – over often many years – they accumulate in thick layers. Drilling down through these layers and removing a section gives us a slice through time, a window into the past, and allows us to learn how it incrementally became the present. The different layers of sediment, distinguished perhaps by texture or colour, cohesiveness or structure, tell us about the kinds of terrestrial environments where they originated – whether these were wetter or drier than they are today, for instance, or whether it was windier or stormier, or whether people were around burning the vegetation to form charcoal.

In such sediment drill-cores, it is also possible to identify old coastlines. Perhaps the core will pass through an ancient beach or a diagnostic coastal deposit (like a mangrove peat), which contains the remains of things that live only in such places. What scientists commonly look for in ocean floor sediments to pinpoint ancient coastlines are concentrations of shells from species that live only between high- and low-tide levels.[24] In many parts of the world, other diagnostic organisms are foraminifera, many species of which live only within a metre or two of the ocean surface and contribute their hard parts (tests) to adjoining beaches.

Many uniquely coastal (or intertidal) vegetation groups create sediments that may be prominent in sediment cores. Think of mangroves, for example. They are wonderful communal plants, hugely useful to coastal dwellers, both directly and indirectly, yet are also efficient trappers of sediment. Mangrove 'muds' are commonly thick, sticky and unforgettably foul smelling. Similar dense coastal vegetation systems like seagrass beds may also be underlain by distinctive sediments that stand out in ocean-floor sediment cores.

How this works in practice is well illustrated by the Gulf of Carpentaria in northern Australia. If you drill through the sediments on the comparatively shallow ocean floor that separates the modern Gulf from the Arafura Sea (to its west), you invariably find a threefold sequence of sediments. At the deepest level, sitting above much older deposits or bedrock, you find evidence of a terrestrial (dry-land) landscape that was emergent when the ocean surface was much lower during the LGM, and the Gulf was a large freshwater lake. Then, as the sea level rose, so the sediments indicative of terrestrial environments were replaced by those indicative of coastal ones, in which mangroves in particular abounded. As the sea level rose even higher, these diagnostic coastal sediments were replaced by ones more characteristic of the deeper ocean floor. Using radiocarbon dating, the age of the key transition zones in these sediment cores can be calculated.[25]

In warm ocean waters, coral reefs often provide far more compelling and complete records of past sea-level changes than can be obtained from sediment cores. Coral reefs are extraordinary, both on account of their much-studied and photographed surface expressions, but also because of what lies beneath. On reefs, the living corals we see near the ocean surface today grow on the skeletal remains of their dead ancestors, which grew on the remains of their ancestors, and so on and so on. So drilling through a coral reef takes us on a journey back through time, sometimes for tens of thousands of years. The remains of the corals growing at different periods can be geochemically analysed to tell us something about the temperature and salinity (a proxy for rainfall) of the ocean water in which they were living. Yet more importantly, coral cores can provide us with datable samples that allow us to calibrate the rise of postglacial sea level.[26]

Our knowledge of the history of postglacial sea-level changes around the Australian continental margin gives us the opportunity to assign ages to each group of Aboriginal stories recounted in Chapter 3. How do we do this? Consider that each story describes a time when the coastline in a particular place was further seaward than it is today. The ancient coastline – its form and even its character – can be reconstructed from information in the story. More importantly, the least amount by which the sea level must have been lower for the story to be true – for this ancient geography to have been truthfully observed by people – can itself be calculated. Thus we can plot the depth of this ancient sea level on the graph in Figure 4.3 and acquire an age range for that observation. Note that the age range is a minimum – the most recent time in Australian history at which observations of the ancient shoreline could have been observed – but that a particular story might actually have originated earlier.

Using this method, minimum age ranges can be assigned to each set of stories discussed in Chapter 3. Let us provide a few representative examples, starting with the Narungga stories of Spencer Gulf (South Australia). Details in several versions of the stories of the drowning of Spencer Gulf talk about it being entirely dry, occupied by 'lagoons and marshes' before the day when 'the sea broke through'. Perhaps the most parsimonious interpretation of these stories is that they describe a time when the ocean was lapping at the lip of the modern Gulf, approximately 50m (165ft) below today's sea level, in which case the stories must be between 11,200 and 12,460 years old. But of course, with our surveys and our satellites, we are biased when it comes to defining landforms like Spencer Gulf; today we can readily see its limits, but perhaps the people who witnessed its inundation many millennia ago were less spatially prejudiced. Perhaps the Narungga stories describe a time when the ocean had already risen halfway up the modern Gulf, referring instead to the drowning of a coastline 22m (72ft) below the modern one, along the line AB in Figure 3.2, in which case the age range for these stories might be just 9,330–9,700 years ago. This is a formidable achievement of memory however you view it.

Then we might consider the example of the Great Barrier Reef off the Queensland coast of north-east Australia, about which 'many tribes ... have stories recounting how the shore-line was once some miles further out ... where the barrier reef now stands'.[27] In the vicinity of Cairns there are numerous stories recalling a time when the coral reef was dry land, even 'all scrubland', and readily accessed by the area's inhabitants. This observation may mean that the contemporary sea level was 40–50m (130–165ft) lower; if the shoreline was where the edge of the barrier reef now stands, then perhaps the sea level was 65m (210ft) lower. Possibly younger stories, like those recalling a time when offshore Fitzroy Island (*gabar*) was a mainland promontory, could have been true when the sea level was 30m (100ft) or more lower than it is today. As determined from Figure 4.3, the minimum ages for such stories range from 9,960 years ago (the most recent possible time at which the sea level was 30m lower than it is today) to a mind-blowing 13,310 years ago (the oldest time within the postglacial period at which the sea level was 65m lower than it is today).

A final example comes from the islands (Rottnest, Carnac and Garden) that lie off the mouth of the Swan River (Western Australia). George Fletcher Moore's colourful account recalls an Aboriginal story about a time when the land separating these islands from the mainland was 'thickly covered with trees' before 'the sea rushed in' and cut them off. As noted in Chapter 3, it would be possible for people taking a meandering path to have accessed these islands from the mainland were the sea level just 5m (16ft) lower than it is today. Yet a sea level 10m (33ft) lower would better satisfy the condition that the area was forested. The lower limit of the age range for the former is 7,450 years ago, the upper limit for the latter 9,140 years ago.

By taking the observations in all the 21 groups of Aboriginal stories recounted in the last chapter and comparing these observations to the sea-level curve in Figure 4.3 using the method described, we come up with minimum age ranges for all. The results are summarised in Table 4.1 below.

As can be seen, while two of these groups of stories are adjudged likeliest to be recent, because they do not demand a lower-than-present sea level, the other 19 are all more than 7,250 years old. At least two – from Kangaroo Island and Cape Chatham – appear to have survived for more than 10,000 years to reach us today.

When confronted by this conclusion, the author's first reaction was scepticism – profound scepticism. For who among scientists, especially conventionally trained scientists inculcated with the dangers of exaggeration and overstatement, would not look askance at a claim of an oral tradition with a longevity of more than 10 millennia? Yet, notwithstanding the caveats of their interpretation, that is what the data show – and no competent scientist should shy away from stating such conclusions.

While it is possible to argue (as in Chapter 2) that for perhaps 65,000 years Australian Aboriginal cultures have been uniquely configured to capture such observations in stories and, more importantly, to render their effective trans-generational transmission, it is possible that similarly ancient

Table 4.1 *Water depths and age ranges for the 21 groups of Aboriginal stories. Age ranges refer to the most recent time at which the observations of lower-than-present water depths could have been made.*

Story number	Story location	Water depth relative to modern sea level (metres)		Age (years ago)	
		Minimum	Maximum	Minimum	Maximum
1	Spencer Gulf	22	50	9330	12,460
2	Kangaroo Island	32	35	10,080	10,950
3	MacDonnell Bay	15	50	8590	12,460
4	Port Phillip Bay	8	12	7820	9300
5	Gippsland	20	50	9240	12,460
6	Botany Bay and Georges River	9	16	7930	9560
	Jibbon Head	5	20	7450	9680
7	Moreton and North Stradbroke Islands	2	8	recent	
8	Hinchinbrook and Palm Islands	5	22	7450	9680
9	Cairns - Great Barrier Reef	30	65	9960	13,310
10	Wellesley Islands (Gulf of Carpentaria)	5	10	7450	9140
11	Elcho Island	5	10	7450	9140
12	Goulburn Islands	17	20	8890	9680
13	Cape Don (Cobourg Peninsula)	9	15	7930	9520
14	Bathurst and Melville Islands	12	20	8210	9680
15	Brue Reef	4	10	7250	9140
16	Rottnest, Carnac and Garden Islands	5	10	7450	9140
17	Cape Chatham	55	60	11,730	13,070
18	Oyster Harbour	4	8	7250	8870
19	Bremer Bay	0	3	recent	
20	Nullarbor Plain near Eucla	10	50	7990	12,460
21	Fowler's Bay	10	50	7990	12,460

stories exist in other cultures. Perhaps these stories have not proved so readily identifiable because they have been watered down by cultural mixing, and the influences of another race's worldviews that have discouraged their preservation. Perhaps they have been dressed up in bearskins of narrative beneath which it is difficult to burrow. Perhaps the languages in which the original stories were recorded and transmitted have died, dragging with them into the graveyard of history all the cognitive paraphernalia that defined a particular group of people. Nonetheless, possibilities remain.

The next chapter therefore moves beyond Australia to see whether drowning stories dating from the period of postglacial sea-level rise might plausibly survive in other parts of the world, allowing their cultures to claim a cultural continuity comparable to that of Aboriginal Australia.

CHAPTER FIVE

Other Oral Archives of Ancient Coastal Drowning

When Nantes-based artist Marjorie Le Berre was growing up in rural Brittany, France, in the 1970s, she once witnessed the activities of a *conteur* or traditional storyteller when he visited a farm close to Les Monts d'Arrée. He was an elderly man, probably in his seventies, and he came to the farm in fulfilment of his traditional role, to tell the people living in this area some of the old stories that he knew, which had been passed down to him by his ancestors, a line of hereditary *conteurs*. Marjorie's father and grandfather had had similar experiences of listening to *conteurs*, who play a traditional role in Brittany society; their stories are only ever told orally, never written down by them. Her grandfather remembered that when he was young, an itinerant *conteur* would be honoured by any village in which he arrived; its inhabitants would typically gather outdoors around a fire in the evening to listen respectfully to his stories, which were enlivened by performance.

One story often repeated by Breton *conteurs* and *conteuses* was about the drowned city of Ys (Caer Ys), which once stood on dry land within what is now Douarnenez Bay, close to the modern village of Camaret-sur-Mer. Prefacing the telling of this story, the *conteur* would recall how today, when the wind blows and agitates the waters of the bay, the bells in the drowned church steeple toll – a constant reminder of the presence of Ys here. Then the *conteur* would relate the story of King Gradlon of Ys, whose daughter did not obey his instructions about the security of the city and, opening its gates during a storm, let in the ocean, drowning Ys forever (see colour plate section).[1]

This story has, of course, been written down on many occasions and in different versions. Various sites off the coast of Brittany for the drowned city have been mooted, which may suggest that a similar series of events happened at more than one location.[2] For example, in the sixteenth century the city of Ys was believed by one writer to have been near Quimper:

> *Still today [in AD 1588] the local people point out the ruins and*
> *remains of the [city's] walls, so well mortared that the sea has not*
> *been able to carry them away, and they say that King Gradlon*
> *was in it at the time it was ruined. These are accidents that have*
> *often happened elsewhere by similar encroachments of the sea ...*
> *but of those things there is no witness but an old rumour noised*
> *from person to person.[3]*

In a well-regarded 1899 account by Paul Sébillot, the town of Is (Ys) is said to have existed once off the coast of modern Saint-Malo, between it and the island of Cézembre, and was protected from the ocean by a dyke. But conflict erupted between the people of Is and those living in the forests of Corseul, inland of Saint-Malo. The king of Is gave the Corseulois a final warning, but they responded by cutting the dyke, making a massive hole through which the sea poured into the town: 'the city of Is was submerged and almost all its inhabitants perished'.[4] Further east, Sébillot recorded that

> *... the fishermen of Cancale say that when the sea is beautiful*
> *and clear, one sees between the Mont Saint-Michel and the*
> *Chausey Islands the debris of walls. They are the remains of a*
> *vanished city.[5]*

Slightly earlier, a prominent Breton abbot, François Manet, made a systematic collection of stories of coastal submergence along the northern coast of France focused, unsurprisingly, on the sites of former monasteries. He identified four of these – Menden, Taurac, Maudan and St Moack's – the last having been burnt 'in a fit of passion' in the year AD 709 by Rivallon, the King of Brittany's brother, who 'afterwards, being penitent, ... re-established it in a better state'.[6]

In the majority of the Breton accounts, the key details involve a town (or monastery) located on the Brittany coast at a time when this lay several kilometres seawards of its current position. It is reasonable to infer that the town's location had become, perhaps for several centuries, progressively threatened by the ocean to the extent that its inhabitants had constructed artificial structures to keep it out. The town's destruction came when this infrastructure finally proved unequal to the task for which it had been designed and the ocean poured into the city, flooding it probably more rapidly than the time it would have taken for everyone to leave – perhaps drowning many people. As time went on, the town was abandoned and the sea rose over it, obscuring it from view, perhaps even leading to its precise location becoming uncertain today.

Such a succession of events – the building of a coastal town, the construction of infrastructure to keep out the ocean, the flooding of the city and its eventual submergence – is plausibly explained as an effect of rising sea levels similar to the observations of Aboriginal Australians related in Chapter 3. The key difference here is that while the sea level reached its present level around Australia some 7,000 years ago, not generally rising more than a metre or two above it subsequently, this is not the case for Brittany, nor indeed for many other Atlantic coasts in Europe. In these places, the sea level has been rising pretty much continuously since the end of the Last Glacial Maximum – the coldest time of the last ice age – 18,000 years ago. Thus any stories about coastal drowning in this part of the world could be significantly younger than has been deduced for those in Australia.

Following the methods for assigning minimum ages to such oral traditions described at the end of the last chapter, we could apply the same to the Ys stories. Yet with these, we also have some historical detail that can be compared with our age estimates, for King Gradlon and others mentioned in some of the Breton drowning stories are believed to have been real people, historical characters whose exploits and their timing are a matter of record. For example, a biography of St Guenolé, who famously drowned the perfidious Dahut, daughter of

King Gradlon, to right the wrong she had done that led to the flooding of Ys, was written in the ninth century and records that he died on 3 March in the year AD 532 at the age of 72 at Landévennec monastery, which he had founded decades earlier.[7] Clearly, if he was indeed involved in the story of Ys, then its drowning can be bracketed in time between about AD 490 and AD 530, just over 1,500 years ago. Yet events far back in history are subject to being concertinaed, transposed on one another even though they were actually separated by centuries, perhaps even millennia. Similarly people, particularly noted characters like King Gradlon, can find themselves having lived numerous lives, apparently involved in incidents centuries, sometimes millennia, apart. So perhaps sea level is a more reliable key to the antiquity of such traditions.

During the last ice age, sea level worldwide averaged 120m (400ft) or so below its present mean level. In north-west Europe, one of the clearest differences in the landscape would have been the connection of the British Isles to the European mainland. They were not as they are today, and have been for most of their written history, an island kingdom, but were then – some 20,000 years ago – an appendage of the continent. One of the last-surviving pieces of the land bridge – Doggerland – was discussed in Chapter 4, but the situation was also quite different further south and west, in the region between Brittany (France) and Cornwall (England). Today the French call the water gap La Manche, while their northern neighbours call it the English Channel, but during the last ice age it was all dry land, readily traversed and almost certainly comparatively densely populated. A river meandered westwards along its axis, cutting a broad valley and eventually emptying into the Atlantic Ocean across the continental shelf (Figure 5.1). Then, beginning about 18,000 years ago, the sea level started rising. Ponderously yet inexorably, the region's geography was transformed. A benign terrestrial landscape was converted to a formidable marine gap, with the few remaining islands generally close to the continent's margins on each shore.

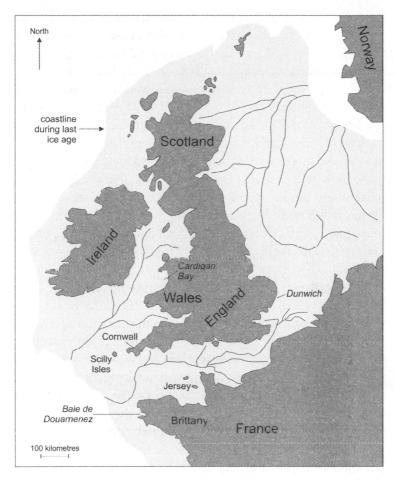

Figure 5.1 *The coasts of north-west Europe today and at the coldest time of the last ice age about 18,000 years ago. The locations of some of the main places discussed are shown.*

Might the stories about submerged cities off the coast of Brittany be distant memories of the postglacial sea-level rise that created today's La Manche? Are there stories from the other side, from the coasts of the British Isles and Ireland, which refer to the English Channel and can be similarly interpreted?

The earlier analysis of the Breton stories – the transition from a coastal town/city that came to need protection from the sea, and its subsequent drowning and submergence – is suggestive of the progressive effects of rising sea levels. As is the case with stories discussed later in this chapter, the drowning of such cities is often recalled as a catastrophic event, something that abruptly destroyed the habitability of such vulnerable coastal places. This is readily explained by the superimposition of extreme sea-level events (like storm surges or even tsunamis) on a steadily rising sea level. The effects of extreme events become incrementally greater even though their magnitude (relative to the average sea level at the time) may not differ significantly from earlier such events. The situation is similar to that of today. The global sea level is rising, but along many coasts it is not the effects of rising mean sea level that are having the most impact. Instead, it is the effect of the extreme events superimposed on this rising sea level that are proving most memorable.[8]

Even though the Breton stories about Ys and other submerged towns and cities might conceivably recall postglacial sea-level rise, there is a clear disparity – of perhaps several millennia – between the time of the observations on which these stories must be based and the time of their historical associations. Yet, as with the 21 groups of Australian Aboriginal stories (Chapter 3), most of which are saying the same thing, the existence of stories beyond Brittany from elsewhere along the coasts of north-west Europe, which all say essentially the same thing, provides the strongest evidence that they also represent ancient human memories of the effects of postglacial sea-level rise. For if they recalled only localised events, then why would they be saying the same thing? Here we review stories from the Channel Islands (off the coasts of northern France), Cornwall, Wales, Ireland and East Anglia, where – to greater or lesser degrees – there are extant stories similar to those from Brittany. Considered together, they are compelling evidence that the indigenous peoples of north-west Europe witnessed postglacial sea-level rise and enshrined their observations in stories that have reached us today, largely through oral means.

The **Channel Islands** (Îles de la Manche) comprise four large islands – Alderney, Guernsey, Jersey and Sark – and a number of smaller ones, close to the coast of northern France. One story, published in 1899, recorded how 'in past times, the [English] Channel was not as great as it is now; one could go to Jersey [from mainland France] without encountering any obstacle other than a brook which was not very wide'.[9] Many of the smallest, now uninhabitable islands in the Channel Islands show abundant signs of having formerly sustained larger numbers of people, presumably when the islands were themselves larger, as would have been the case when the sea level was lower. The names of some of these islands, particularly those like Ecréhous, Brecqhou and Lihou, which include the suffix *hou* (meaning house), suggest that they were once large enough to have supported sizeable populations. Yet the main evidence of once greater populations comes from thick, dense shell middens on some of these tiny islands, thought to have accumulated on the islands when they were larger and inhabited by numerous shellfish-eating people.

Submerged forests are a tangible, although not always readily observable, feature of the shallow ocean floor surrounding these islands and separating them from the adjacent continental coasts. In 1787, a local newspaper on Jersey commented that:

> *The trunks and roots of trees which showed themselves last winter by the agitation of the sea in St. Ouen's Bay, and which are still visible, furnish us with a subject of contemplation relating to times very remote. One sees thousands of trees laid one close to another in this bay …*[10]

Today, the ocean floor between Jersey and the nearest parts of the French mainland reaches depths of 10–12m (33–40ft).

Crossing the English Channel, the people of **Cornwall** in south-west England (including the offshore Scilly Isles) have some stories about drowned coastal lands, perhaps the most famous of which is Lyonesse.[11]

Lyonesse has become grounded in English legend as the rumoured site of Avalon, to which King Arthur was carried to die, yet it also appears to be anchored in history. One of the earliest references pinpoints the date of its submergence to 11 November 1099, when it is said that 'the sea overflowed the shore, destroying towns and drowning many persons and innumerable oxen and sheep'.[12] Much later, in *A Tour through Cornwall in the Autumn of 1808*, the Reverend Richard Warner recounts that:

> *William of Worcester ... states, with a degree of positive exactness, stamping authenticity upon its recital, that between Mount's Bay [near Land's End] and the Scilly Islands there had been woods, and meadows, and arable lands, and a hundred and forty parish churches, which before his time were submerged by the ocean.*[13]

If Lyonesse ever existed, and there are many who regard the story of it as allegorical, then it is possible that it was submerged by postglacial sea-level rise. Perhaps, as might have been the case for Ys, sea-level rise served simply to amplify extreme wave events, one of which tore through Lyonesse and forced its abandonment; subsequent sea-level rise hid it from history. The big wave may have been a tsunami caused by an ocean-floor earthquake in the vicinity. It is a long stretch, but one story about the destruction of Lyonesse recalls that one of its prominent residents, Trevilian, succeeded in saving his family after the first wave hit, but then barely managed to save himself after the second (or third) wave washed over him and his horse. This wave was reportedly the highest of a series, similar to the tsunami wave trains that often characterise such events. Sometimes the earthquake not only generates a wave train but also causes islands close to the epicentre to sink rapidly,[14] and it is possible that an island called Lyonesse was abruptly submerged during an earthquake off the coast of modern Cornwall. Tsunamis do affect this area. The great Lisbon Earthquake of 1 November 1755 that destroyed the Portuguese capital drove a train of tsunami waves onto coasts around south-west England; the third wave was reportedly by far the highest.[15]

When visited by Romans, perhaps around the year AD 10, the Scilly Isles are said to have numbered just 10, but by 1753 there were 140 of them. This could be interpreted as evidence for sea-level rise within this period – larger islands being subdivided by inundation – although it would be imprudent to place too much faith on the precision of the Roman count.[16] Yet there is considerable evidence from the Scilly Isles of their submergence within the past millennium or more. It includes traditions of once-contiguous islands, and the remains of stone walls (locally called 'hedges'), former field boundaries running from one island to another along the sea floor now covered by 3–4m (10–13ft) of water.[17] Together with the Lyonesse tradition, such evidence for submergence does point to an observed history of sea-level rise comparable to that inferred for Brittany … and to points north.

Many traditions about 'lost cities' and 'sunken palaces' refer to the area off the west coast of **Wales**, particularly in Cardigan Bay, where perhaps 'sixteen cities' of Cantre'r Gwaelod are reputed to have been submerged. The similarities between the extant stories of the submergence of these cities and those in Brittany have been remarked upon and suggest that – at least at some point – traditions fused. The question remains whether there was in fact any original story in either Breton or Welsh tradition that, taken in tandem with empirical evidence for sea-level rise, suggests that people in these places observed the effects of rising sea levels … and that their observations still survive. The review of the Breton traditions above implies that they did, but what about the Welsh?

In most Welsh stories about Cantre'r Gwaelod, the king is Gwyddno (rather than Brittany's Gradlon), and it is his intemperate steward Seithenhin (rather than a dastardly daughter) who unlocks the flood gates causing the city to be drowned.[18] The city, often said to be one of many, dominated a massive area of low ground named Maes Gwyddno (the

Plain of Gwyddno), perhaps more than 2,000km² (770mi²), in an area between Cardigan and Bardsey Island and the shores of modern Cardigan Bay.[19] This land was said to have been 'extremely good and fruitful and flat', and its loss must have had a memorable impact on its occupants; an account from about AD 1450 mentions 'the lament of Gwyddno ... over whose land God turned the sea'.[20]

While some scientists have railed against any literal interpretation of these Welsh stories, rightly so in some of their more extravagant permutations, others have been persuaded that the story of Cantre'r Gwaelod is more likely 'a legendary account' dating from at least the fifth or sixth century 'of an actual event'.[21] As was argued for Australian Aboriginal societies (Chapter 2), 'the conservatism of this tradition in both Wales and Brittany ensured the survival of this tale to the present day'. This analysis does not compel us to believe that these stories were based on observations in the fifth or sixth century, but rather that older stories about 'great inundations' became attached in medieval times to the heroes of the fifth and sixth centuries, people credited with founding Welsh (and perhaps Cornish and Breton) traditions.[22] Similar stories about the Welsh coast that scholars have interpreted as built around memories of coastal inundation include that about Tyno Helig ([Lord] Helig's Valley), which may have covered some 46km² (18mi²) of land now underwater in Conway Bay,[23] and the town of Caer Arianrhod in Caernarvon Bay, the ruins of which are reported to sometimes emerge at low tide.[24]

In the case of Wales we do not have to depend solely on such echoes of ancient traditions, for there is an abundance of physical evidence of former shorelines in Cardigan Bay and elsewhere by which scientists have been able to track the progress of the postglacial sea-level rise that flooded the area. There are submerged forests within a kilometre or two of the modern coast of Cardigan Bay (see colour plate section), but further out there are sediments containing peats that mark a time when forests of oak and pine stretched across much of this shallow bay.[25] Were the sea level around 30m (100ft) below its present level, much of Cardigan Bay – at least that between Cardigan and Bardsey – would be dry land, a condition that would have been met at least 9,000 years ago. Given that the present form of Cardigan Bay was established a thousand years or so later, it is difficult to imagine a sizeable land mass being emergent and connected to the mainland more recently.

During the Last Glacial Maximum, about 20,000 years ago, when the sea level in this part of the world lay some 125m (410ft) or so below its present level, **Ireland** was connected to Wales and England, which were in turn connected to what is today mainland Europe. Postglacial sea-level rise gradually drowned the Ireland–Wales land bridge, leading to its submergence about 9,600 years ago, or perhaps a few millennia earlier.[26] The timing is important, for it has the potential to lend credence to or to dismiss the idea that the traditions of unassisted human crossing of the Irish Sea may be based on actual events. The most common story is that of Brân the Blessed (Bendigaidfran), a Welsh hero, who went with his warriors to Ireland to rescue his sister from a bad marriage. In one account, at a time when the Irish Sea was 'not so wide', Brân crossed at least part of it 'by wading', but thereafter 'the deep water grew wider when the deep overflowed the kingdom'.[27] The story is reminiscent of Aboriginal Australians once having crossed places like Backstairs Passage or Clarence Strait by 'a combination of walking and wading', something that has not been possible for several thousand years (see Chapter 3). So might a similar explanation apply to the Irish Sea at the time of Brân? It may, but there are currently too many uncertainties to consider this likely, although one can see how, to rationalise the crossing of what is now a deep sea, it is necessary for Brân to be a giant in most extant versions of this story.[28]

What, therefore, can we say about the possible antiquity of stories of coastal drowning from the Atlantic coasts of north-west Europe? Using the same methods as for the Australian stories, explained at the end of the last chapter, we start by estimating the minimum depths below present sea level at which each group of stories (clustered by location) would be true. These depths are shown in Figure 5.2 below, together with sea-level histories for Brittany, Cornwall and Wales that cover the later part of the period of postglacial sea-level rise.

The depth minima on the right of Figure 5.2 are of course estimates – the stories are too vague for anything else – but considered realistic nonetheless. Thus, for example, a submerged town (like Ys) a kilometre or so off the modern coast of the Baie de Douarnenez would today be 15–30m (50–100ft) below sea level. And if Cantre'r Gwaelod existed somewhere between 5 and 20km (3–12 miles) off the modern coast of Cardigan Bay, then it would today lie beneath 10–20m (33–66ft) of water. These kinds of numbers are not intended to be definitive, beyond disputation, but rather

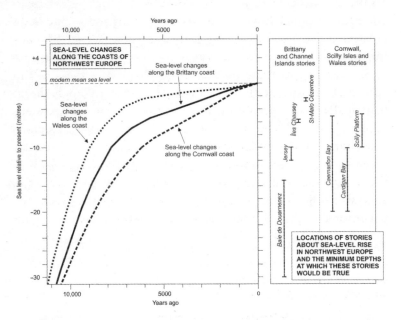

Figure 5.2 *Sea-level changes along the coasts of north-west Europe over the past 10,000 years or so are shown on the left. On the right is shown a selection of locations of compelling stories about sea-level rise from north-west Europe, represented by bars showing the minimum depths at which these stories would be true.*

realistic approximations of where submerged places may once have been – a necessary preliminary to determining the minimum ages of these stories.

By relating the minimum depths shown on the right of Figure 5.2 to the appropriate sea-level curve on its left, minimum ages for these groups of stories can be calculated. For instance, by drawing horizontal lines from the top and bottom of the depth range for stories from the Baie de Douarnenez to the Brittany sea-level curve, then vertical lines from the points of intersection to the age scale, we can determine that for these stories to be true – to be a distant recollection of something people actually observed here – they must have endured for a minimum of 8,750–10,650 years. Similarly, for the stories from Cardigan Bay to be true, assuming the sea level at the time of observation to have been 10–20m (33–66ft) lower than it is today, they must have been created at least 9,000–10,250 years ago. The full set of dates for these stories is shown in Table 5.1.

Similar caveats that were identified for the analysis of the Australian stories at the end of the last chapter apply here, principally those involving the crudity of both the water-depth data and the uncertainties in the stories. Before proceeding, mention needs to made of Lyonesse, which in several accounts is said to have once occupied the area *between* the Scilly Isles and Cornwall now covered by ocean. In the analysis below (Table 5.1), it is assumed – solely for reasons of parsimony – that the stories about Lyonesse recall a time

Table 5.1 *Water depths and age ranges for the seven groups of sea-level rise stories from the Atlantic coasts of north-west Europe. Age ranges refer to the most recent time at which the observations of lower-than-present water depths could have been made.*

Story location	Water depth relative to modern sea level (metres)		Age (years ago)	
	Minimum	Maximum	Minimum	Maximum
Baie de Douarnenez	15	30	8,750	10,650
Jersey (to mainland France)	10	12	7,800	8,200
Îles Chausey	5.5	6	5,800	6,200
St-Malo Cézembre	2.5	3	2,850	3,400
Caernarfon Bay	5	20	7,700	10,250
Cardigan Bay	10	20	9,000	10,250
Scilly Isles (platform)	5	10	7,700	9,000

when the shallow platform from which the Scilly Isles rise today was dry land; in other words, Lyonesse was Scilly. This is assumed because the ocean floor between the Scilly Islands and Cornwall is today comparatively deep. A narrow, sinuous dry-land connection may have been possible when the ocean surface was 65m (210ft) lower, but for this to have been a few kilometres wide, it would have to have been 70m (230ft) lower. Although these depths are not shown in Figure 5.2, it is likely that the minimum age for the narrow land bridge would have been around 14,600 years ago, that for the wider one a millennium or so earlier.[29]

There are other drowning stories in Europe that do not lend themselves to being dated in this way but are still worth mentioning.

The east coast of England defines the western limit of the North Sea, the county of **East Anglia** marking the western stepping-off point for Doggerland, the broad land bridge that connected the British Isles to mainland Europe during the last ice age. The rise of the sea level in postglacial times that saw the dismemberment and finally the drowning of Doggerland has also had dramatic effects on the poorly consolidated (and easily eroded) sediments that comprise much of the East Anglian coast. When you visit the coastal village of Dunwich (population about 183) today, it is astonishing to consider that it was once among the 18 largest towns in England, sitting at the head of a sheltered natural harbour, the bustling hub of a vibrant trade in fish, salt and cloth.[30] However, well before it reached its zenith, Dunwich had already started losing its battle with the sea. Long affected by shoreline erosion (marked by cliff retreat), some 600 houses are reported to have been washed offshore during the winter storms of AD 1287–1288. Since then this sad story has continued, involving a loss of status, abandonment by merchants and a continuing physical loss; the last piece of All Saints, the only church to remain standing after the AD 1740 storms, collapsed into the sea in AD 1922 (see colour plate section).

The sedimentary vulnerability of the East Anglia coast is amplified by land sinking, one expression here of the slow, enduring downwarping of the North Sea Basin. Not only is this basin sinking − like many similar basins − under the growing weight of the sediment accumulated on its watery floor over millions of years, but it is also being undermined by the flow of its underlying crust to areas of surface rise, for example in Scandinavia.[31]

Given such a situation, you might expect East Anglian traditions to be rife with stories of 'sunken cities' and the like, but they are not. Aside from those that insist − as in countless other places with similar histories − that the droll sound of tolling bells can be heard from beneath the sea's surface when it becomes agitated, I know of no other known folk memories of what was a prolonged and memorable process for the region's inhabitants. This may be because, unlike in most other places from which we have such 'drowning stories', East Anglia has been substantially repopulated − quite frequently, perhaps, during the past few millennia − as successive waves of migrants and invaders, from Vikings to Normans, overran the country. Traditions would have been lost, and connections between people and place forgotten, while the vagaries of living along an unstable coast would have been rediscovered anew on many occasions.

Along the shores of the **Mediterranean Sea** are numerous 'sunken cities' that are of only passing interest to the topic of this book. For these cities are well known, and many have been fulsomely documented, not least because of the mass of written records that recall their existence ... and sometimes the turmoil that accompanying their disappearance. Many of the cultures associated with them were literate so there is little dispute, if the obvious fictionalisations are sidelined, that they once existed and are now underwater. There is no need to depend in most instances on oral traditions. Add to this the fact that the material culture associated with most submerged Mediterranean cities is generally conspicuous; you cannot readily miss the statuary, the columns and harbours, towers and temples, walls and wells, indeed all the constituents of Greco-Roman civilisation that are well known to travellers and museum-goers in much of modern Europe.

A good example comes from the Nile Delta, where the Greeks built two cities – Herakleion[32] and Eastern Canopus – to facilitate trade with Egypt and other parts of North Africa around 2,000 years ago. Once located on the now-infilled Canopic distributary channel of the Nile, the remains of these cities currently lie 5–7m (16–23ft) below sea level. Some 60 per cent of the sinking these cities have experienced is attributed to a combination of rising sea levels and the slow yet inexorable compaction of the delta sediments on which they were built. The remaining 40 per cent or so happened more abruptly as a result of the episodic collapse of the water-soaked sediments on which the cities were stood.[33] Owing to the abundant written records about and from these cities, they were never truly 'lost' to humanity, although until about 1999 no one was quite sure exactly where they had been. Deltas are dynamic places, routinely flooded, with waters moving the loose sediments from which they are built from place to place; new distributary channels are carved during floods while others may be come infilled. Physical collapse often occurs, particularly along a delta's seaward fringes, where underwater slopes are generally steepest. And of course in the eastern Mediterranean, which is more affected by tectonic paroxysms than most places, the impact of nearby earthquakes and locally generated tsunamis has often been to destabilise places like the Nile Delta.[34]

The seismo-tectonic and volcanic activity that affects much of the Mediterranean is another reason why this region's submerged cities do not get much of a mention in this book. The land moves up and down – sometimes slowly, occasionally abruptly – as a result of the converging African and Eurasian plates. The former pushes northwards, while the latter resists but of course periodically fails, so in places each is sliced up and thrust beneath its neighbour. Friction causes earthquakes and oftentimes these cause the abrupt – and catastrophic – sinking of coastal lands. Take the Greek city of Helike, once the principal city in the district of Achaea, which is recorded as having been comprehensively destroyed during an earthquake (and accompanying tsunamis) in 373 BC. In fact, so comprehensive was its destruction that for centuries no one was quite sure where Helike had actually been ... or indeed whether it had ever been! The basic problem was that the earthquake, which occurred along a thrust fault associated with the movement of the African Plate beneath the Eurasian, had caused Helike to sink by 3m (10ft) within a very short period of time.[35] Then the tsunami came and dumped a layer of sediments over the top and thereafter, because the geography of the area had been comprehensively reconfigured, the site of Helike found itself within an ocean passage, and subsequently became buried by river and ocean sediments. It took some extraordinarily persistent detective work to relocate Helike, which today, because of all this sedimentation, is now back on the land – where almost nobody had thought to look for a sunken city.[36]

A few hundred kilometres away from these thrust belts, perhaps the same distance below the surface, the melting of downthrust crust produces masses of liquid rock that often finds its way to the ground surface, producing lines of active volcanoes in places like the Aegean Volcanic Arc (along which Santorini famously erupted, perhaps inspiring Plato to invent Atlantis, in 360 BC) and the Calabrian Arc (responsible for a range of volcanic activity from Etna to Vesuvius and beyond). A good example from the latter region of how volcanic activity can cause coasts to sink concerns the Roman-era ports of Baia, Miseno and Pozzuoli, parts of all of which now languish several metres below sea level. These former ports are all located within a caldera of the Phlegrean Fields volcanic complex (Campi Flegrei), which is notoriously active today. They were repeatedly submerged and uplifted on at least three occasions since their Roman occupation (see colour plate section).[37]

European and North African coasts do indeed have many submerged cities associated with them. Yet *both* the submergence of most of these is ascribable to processes other than postglacial sea-level rise (so submergence is consequently more recent than most of the other examples described in this book), *and* written records exist for most, removing any need to rely on oral traditions (which are the main source for the 'recollection' of stories about the other examples in this book). We now therefore shift our attention east to the coasts of the Indian subcontinent, where drowning stories more akin to those of Australia may exist. But before we describe these stories, consider that parts of the coast of **India** have been affected more recently by some epic subsidence.

On 17 December 1846, the members of the Geological Society of London assembled for their quarterly meeting. The title of one of the items, forwarded from a correspondent in India and read to the meeting – *Extract from a Letter concerning a Depression lately produced in consequence of an Earthquake in Cutch, by Mrs Derinzy* – puts one in mind of a nervous memsahib suffering regular attacks of the vapours, who experienced a minor earth tremor and suffered a breakdown as a consequence. But it is not that kind of depression that Mrs Derinzy wrote about. The depression was that of the ground surface.

The Rann of Kachchh (Cutch or Kutch) is a low-lying, shallow basin, fault bounded, and close to the modern border between India and Pakistan. It is unclear how long this *rann* (salt marsh) has existed, for in medieval times the area where it is now was connected to the Arabian Sea by a prominent distributary of the Indus River named the Kori. Today the Kori is no more, a natural dam 80km (50 miles) in length having blocked it. This dam, named Allah Bund, was created during the earthquake that rocked the area on 16 June 1819. The uplift that created Allah Bund was complemented by subsidence, which dropped an area at least 7km (4 miles) in diameter (including the British fort at Sindri) more than a metre.[38] The resulting landscape change was even more impressive, for the combination of the uplift and the subsidence created a basin with a surface area of more than 1,000km^2 (385m^2) – the Rann of Kachchh.

Earthquakes affected the same area in the 1840s; Mrs Derinzy's Depression probably occurred in the course of the one in June 1845. During this event, as in most, the earthquake was followed by a tsunami that flooded the low-lying Rann.

Subsidence also occurred. An extract from Mrs Derinzy's letter talks of a man crossing the area on a camel, expecting it to be dry. Instead,

> ... the guide travelled 20 miles [32km] through water ... up to the beast's body. Of Lak [Lakhpat] nothing was above water but a Fakeer's pole (the flagstaff always erected by the tomb of some holy man); and of Veyre and other villages only the remains of a few houses were to be seen.[39]

The subsidence experienced in the Rann during the 1840s served to extend its lowland area, but it also raised the Allah Bund to new heights, creating a more effective dam for the Kori and ensuring that the landscape changes enacted in 1819 would endure.

Differential tectonics have affected the area now occupied by the Rann of Kachchh for probably millennia, but details of the earliest changes that the people of this area witnessed have probably been lost forever. Yet maybe they contributed, along with countless other similar examples, to oral traditions that eventually became formalised when literacy reached India more than 3,000 years ago, and the earliest and still some of the most revered books about the history of the subcontinent were produced. For example, the poems of the *Rig Veda*, perhaps the earliest extant Sanskrit text, include one about Varuna, the Hindu god of water and the ocean, who 'led the watery floods of rivers onward ... [and] made great channels for the days to follow'. The poem talks about when 'he sinks in Sindhu ... ruling in depths and meting out the region, great saving power hath he, this world's Controller'. It may not be beyond the bounds of probability to link the tradition of Varuna, who occupies an underwater world, with millennia-old observations of land subsidence along the coast of India.[40]

There is more such material in the *Ramayana* and *Mahabharata*, the Sanskrit sagas considered to have been originally compiled from oral traditions. Three case studies, each deriving from one or both of these texts, which recall the submergence of coastal land, are recounted below: Dwaraka (Gujarat), Ramasetu (Tamil Nadu) and Mahabalipuram (Tamil Nadu) (Figure 5.3). The imprecision of these stories – compared to some described earlier in this book – makes it impossible to reconstruct the depths of sea level to which they might refer, and therefore to calculate minimum ages for the stories using known sea-level history.

The golden city of **Dwaraka** (sometimes Dvaraka or Dwarka), long the abode of Lord Krishna, was somewhere close to the western extremity of the Saurashtra Peninsula, about 250km (155 miles) west of the Rann of Kachchh, and

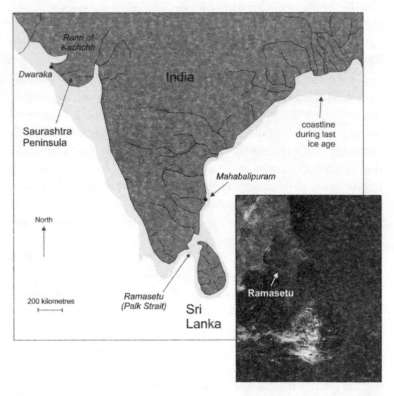

Figure 5.3 *The coasts of India and Sri Lanka today and during the coldest time of the last ice age, about 18,000 years ago. Locations of key places are shown, together with a satellite image showing a shallow Ramasetu. Satellite image credit: Jacques Descloitres, MODIS Land Rapid Response Team, NASA/GSFC.*

well positioned to play a prominent role in former cross–ocean exchange networks, especially within the Arabian Sea to the west.[41] The *Mahabharata* tells that a town existed on the site of Dwaraka before it was built, a place called Kusasthali that was destroyed by the sea. Dwaraka was then constructed on the same part of the coast, including an area of 12 *yojana* reclaimed (or needing to be protected) from the sea.[42] Why anyone would want to reclaim land on which to build a city when plenty was available above sea level is unclear.[43] A plausible explanation is that two separate (and significant) stories are conflated here, as often happens with such ancient traditions. The first concerns the construction of a coastal city, the second the growing need – its urgency perhaps amplified by the destructive impacts of an extreme wave event – to protect that city from rising sea levels, a process involving major earthworks along the shoreline. But in the end all this proved insufficient, and shortly after Lord Krishna left the city was submerged … since which time it has become lost, its precise location uncertain. It is possible that extreme waves (storm surges or tsunamis), superimposed on a rising sea level, brought about the end of Dwaraka. A story in the *Mahabharata* states that many residents vacated the city before it was flooded,[44] implying that the submergence was neither sudden nor unexpected. In fact, it may be that this detail conflates the experiences of many generations, and that submergence, perhaps punctuated by extreme short-lived events presaging the 'end', could even have taken 100 years or so to accomplish.[45]

Today, thanks largely to the efforts of Indian underwater archaeologists, we are fairly sure that Dwaraka was located in coastal Okhamandal and included offshore Bet Dwarka Island. At the latter site, underwater surveys have shown the existence of stone walls – perhaps jetties – and slipways, as well as numerous anchor stones and other artefacts (see colour plate section). Similar features are found on the adjacent mainland around the mouth of the Gomati River. Widely heralded as 'proof' of the stories in the *Mahabharata*, these discoveries do yet pose some difficulties concerning the time when Dwaraka was submerged.

Three dating methods have been employed. The first involved thermoluminescence dating of buried potsherds at Bet Dwarka; the earliest set of dates suggested that these were fired some 2,220–3,870 years ago (an average of about 1100 BC).[46] This result is broadly consistent with that obtained from celestial observations during the lifetime of Lord Krishna that are found in many ancient texts; an age range of 1500–1400 BC has been suggested.[47] But then there have been several suggestions that Dwaraka (or perhaps earlier Kusasthali) existed during the later part of the Harappan Period, perhaps around 2000 BC or even earlier.[48] Whether or not we might confidently ascribe the submergence of Dwaraka to postglacial sea-level rise is dealt with below, after the discussion of the other two sites in India.

The rock bridge of **Ramasetu** (Rama's Bridge), between part of the South-east Indian coast and that of nearby Sri Lanka, is reported in the *Ramayana* as having been constructed by the monkey army of Lord Rama, who used it to cross the gap to rescue his wife, Sita, who had been kidnapped by the demonic king Ravana.[49] The monkeys built the bridge using rocks from the hills and trees ripped up by their roots (see colour plate section). Led by Lord Rama, a huge monkey army then crossed to Sri Lanka, and Sita was eventually restored to him. Astronomical and historical-genealogical dating places the writing of the *Ramayana*, generally thought to pre-date the *Mahabharata*, some time between 4000 and 2000 BC.[50]

When the sea level was lower during the last ice age, there would have been a broad land bridge in the area connecting what are now India and Sri Lanka. As the sea level rose, so the land bridge would have gradually been reduced in size and eventually drowned, undoubtedly to the consternation of people living at its extremities. Perhaps, as elsewhere in the world, memories of the time when a land connection between India and Sri Lanka existed found their way into folk traditions that in turn informed the stories in the *Ramayana*. Tens of generations after such a bridge existed, how would people ever explain that it had? Perhaps they melded the story with religious beliefs and culture heroes, massaging their reported antics to accommodate the story, resulting in the memory of Ramasetu being preserved.

One point of interest to geologists is that in some stories about Ramasetu, before Rama's monkey army built its bridge, there had been another: one that ultimately failed.[51] Might this represent the memory of a rise of sea level that drowned the land connection between India and Sri Lanka and was followed – perhaps hundreds of years later – by a fall of sea level that once again exposed it? Then came its final submergence, leading to the situation we see today where an underwater trace of such a connection is visible on satellite photos (see inset in Figure 5.3). That may strike you as a bit far-fetched, a *deus ex machina*, but actually from a geological standpoint it is not, for many scientists studying postglacial sea-level rise have found evidence consistent with oscillations – wiggles – rather than a smooth rise. So it may be that people witnessed the submergence of the India–Sri Lanka land bridge on more than one occasion, a detail that has also succeeded in finding its way down to us today.[52]

The final case study refers to the city of **Mahabalipuram** on India's east coast, which has a long history of submergence, so much so that its earliest part once slipped into the realm of legend – once, that is. For shortly before the great Indian Ocean Tsunami struck Mahabalipuram on 26 December 2004, local fisherfolk saw the sea draw back an incredible 500m (1,640ft), exposing on the bare sea floor the remains of temples and associated statuary, numerous pieces of which were subsequently ripped from their shallow sandy graves and dumped inland as the massive waves rushed onshore. In the aftermath of the tsunami, the shoreline here was found to be littered with huge granite blocks (once parts of defensive walls), and statues of lions and elephants that once sat proudly above gates to a Pallava-era port city, perhaps 1,300 years ago.[53] However, Mahabalipuram was an important port far earlier, perhaps as much as 2,000 years ago, and had trade links with Imperial Rome, China, Sri Lanka and of course other parts of India.[54] Subsequently, at least since Marco Polo's time (around AD 1275), Mahabalipuram was known as the City of the Seven Pagodas for its seven famous temples (topped with iconic pagodas, or *kalash*), all but one of which later disappeared beneath the ocean surface. In the year 1776, memories of these were fading, for then only older people

> *... remembered to have seen the tops of several pagodas far out in the sea, which being covered with copper, probably gilt, were particularly visible at sunrise as their shining surface used to reflect the sun's rays.[55]*

A Dutch portolan chart made in AD 1670, named DE CVST VAN MALEBHAER (The Coast of Malabar), is believed to show some temples at Mahabalipuram that were at the time very much onshore. It is possible that some of these later disappeared and perhaps gave rise to − or helped sustain − older stories about submerged temples here.[56] Yet most evidence, including that provided by underwater surveys, suggests that the six vanished temples with their pagodas were submerged at least 1,000 years ago, but perhaps no earlier than AD 600, when the Pallava culture, which seems likeliest to have created these gilded temples, began to flourish at Mahabalipuram.

We now, therefore, turn to look at the history of sea-level changes along the coast of India to try to determine the antiquity of the stories from these three locations. Surprisingly, there has been little effort to utilise the vast amount of local information about the evolution of the coast of India to produce a region-wide picture of sea-level change since the last ice age. In the absence of such information, sea-level histories from Malaysia, Singapore and the offshore islands of the Maldives can be brought together to produce a reasonable approximation of the situation in India. That is not really good enough, you might charge, but remember that its purpose is not to interpret with a very high degree of precision, but only to obtain an approximate measure of how old particular stories might be.

Our synthetic sea-level curve for India involves the sea level rising rapidly after the end of the last ice age to reach its modern level about 6,800 years ago – similar to Australia. It slowed down a bit after that before rising again to a maximum, perhaps 2m (6½ft) higher than it is today, about 5,200 years ago. It then started falling but then once again started rising, reaching another maximum – a bit lower than the earlier one – about 2,350 years ago, after which it fell, probably with some minor oscillations, to its present level (Figure 5.4). What this means is that the drowning stories from India recounted above could recall any one of these three highstands of sea level. The question is which one.

If we accept that Dwaraka – and its predecessor Kusasthali – were indeed Harappan cities, then the most likely scenario is that Kusasthali was established about 3,000–3,500 years ago. Then, as the sea level started rising thereafter, the city began to suffer from the effects of this. A bold new leader – Lord Krishna arose and declared that Kusasthali should be comprehensively redesigned so that it might be protected from the rising sea. This leader oversaw the building of the sea defences recalled in the *Mahabharata*, and the new city, when completed around 2,800 years ago, flourished in the twilight centuries of the Harappan culture, as the focus of this was driven southwards into moister Sausashtra by an increasingly drier climate further north. Yet the sea level continued rising and gradually the golden city succumbed to the waves by about 2,400 years ago.

It is a different situation when it comes to Ramasetu, simply because the *Ramayana*, in which it was first written about, is thought to have been produced some time between about 4,000 and 6,000 years ago, so that the story is likely to pre-date the most recent sea-level maximum. Perhaps we should not place too much trust in that deduction, for ancient books are notoriously difficult to date and of course many evolve; it is even possible that the story about the building of Rama's bridge, today considered such a defining detail of the *Ramayana*, was in fact a more recent addition. But for the sake

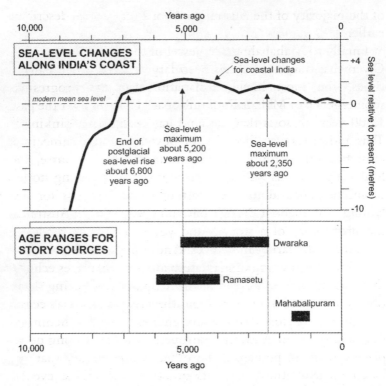

Figure 5.4 *Sea-level changes around India over the past 9,000 years or so. These are represented by an envelope to show the uncertainty based on imprecision of dating, as well as the tidal niches in which particular indicators are found. The sea reached its current level (dashed line) about 6,800 years ago, and stabilised for a while before peaking about 2m (6½ft) higher than it is today about 5,200 years ago. Then it fell, reaching a low about 3,200 years ago, before rising once more to a lower peak about 2,350 years ago, from which it fell more or less to its present level. Likely age ranges for stories about Dwaraka, Ramasetu, and Mahabalipuram are shown as solid bars.*

of argument, assuming that this was *not* the case, then the most parsimonious interpretation is that Ramasetu was last visible when the sea level was lower than it is today some 6,800 years or more ago. This may make it a clear contemporary of the majority of the Australian Aboriginal stories described earlier.

Finally to Mahabalipuram, jewel of the Pallava monarchy. Given that the sea level has been largely stable since Pallava times, the most likely explanation for the progressive submergence of Mahabalipuram's coastal fringes over the past 1,300 years or so is that the land has been slowly sinking.[57] This indicates that the City of Seven Pagodas came into existence only around a millennium ago. Of course, for Mahabalipuram as well as other comparatively young stories of submergence along the coast of India,[58] just as for the 'young' stories about Bremer Bay and Oyster Bay in Australia, the probability of a story being young does not render it immune from earlier influences. And it may well be that the story of the drowning of Ramasetu and similar places echoed through Indian cultures for millennia, influencing later observations and inferences about the history of India's coast.

If drowning stories could have endured many millennia as oral traditions (in Australia, Europe and India, among other places), then so, perhaps, have other types of story that are based on eyewitness accounts of other memorable events. This is the subject of the next chapter.

CHAPTER SIX

What Else Might We Not Realise We Remember?

B ack in the 1990s a greenhorn geologist, intrigued by the theoretical possibility of volcanic eruptions having occurred within the Fiji Islands archipelago during the period of its human occupation of some 3,000 years, determined to investigate a likely candidate for activity – the imposing, rainforest-cloaked mountain named Nabukelevu (meaning 'the big yam mound') at the western extremity of the island of Kadavu. A succession of boats got him to his destination, and the hospitable people of Daveqele village housed and fed him, while he spent several days on the mountainside following tracks made by feral pigs – the only thoroughfares – hoping to stumble across field evidence for recent volcanic activity. The nights he spent in Daveqele, listening to the people's stories, enlivened by drinking *kava* made with the magical water that gushes from springs around the base of Nabukelevu, ultimately proved far more fruitful.

One of the stories told of a god named Tanovo, who resided on Ono Island, 50km (30 miles) north-east of Nabukelevu, and made a habit of sitting on the beach every evening contemplating the sun sinking slowly beneath the horizon. He did this until one day he found the horizon completely changed, besmirched even, by a monstrous protrusion – Nabukelevu – which had suddenly appeared to obstruct his view of the setting sun. In fury, Tanovo flew to western Kadavu to confront Tautaumolau, the upstart god of Nabukelevu. A battle ensued, but Tanovo was eventually bested and flew away, dropping earth as he went at the islands of Dravuni (meaning ash) and Galoa. Could it be, wondered the geologist, that an eruption of Nabukelevu – unarguably a youthful volcano – was witnessed by humans on nearby islands, who encoded their observations in myth so they might be passed on through the centuries to warn their descendants ... and to intrigue outsiders? That geologist still thinks this is the case.[1]

Up to this point the focus of this book has been on memories of coastline drowning – some plausibly recalling postglacial sea-level rise – that have come down to us today through largely oral means. The implication is that some such memories have been successfully passed continuously between hundreds of generations for several thousand years. This deduction is so astonishing yet apparently so unassailable that it seems worthwhile to now consider what other ancient memories we, as a species, might have ... and whether the antiquity of any of them matches those of the drowning stories. The example above is fairly solid evidence for a memory of a volcanic eruption having been preserved for perhaps 2,000–3,000 years. There are, of course, others, including the remarkable Klamath story of the eruption of Mt Mazama (and the associated formation of modern Crater Lake) recounted in Chapter 1, which appears to be around 7,600 years old. But how common are these stories? Are these examples anomalies, predestined for preservation by unique cultural circumstances, or are they representative of a largely undervalued body of knowledge? This chapter attempts to answer this question by looking – with a focus on Australia, where circumstances seem especially favourable – at stories of volcanic eruptions, abrupt land movements, meteorite falls and extinct animals.

Volcanic eruptions provide excellent material for mythmaking: they are frequently dramatic and therefore memorable; often shockingly disruptive and therefore imperative to explain to future generations so that they might recognise the warning signs and know what might follow. Volcano myths are common in the pantheons of collective memory created by many of the world's cultures whose people live or have lived in areas of volcanic activity. Japanese, Italian and Norse cultures, to name but three, are replete with volcano stories – about the fires within the Earth, about how feuding deities rain fire on one another, and about how fire is used as a weapon or as punishment. However, this exemplifies one of the main challenges of identifying the antiquity of volcano myths in such places, namely that across a timespan of perhaps a few thousand years, volcanic activity has occurred so regularly that it is likely an original myth has been many times rejuvenated, and even embellished, by successive observations of similar phenomena.

Such a situation is well illustrated by the highlands of New Guinea, where legends about ash falls – often celebrated *post facto* for the fertility they restore to overworked soils – proved almost impossible to link to particular eruptions or even particular volcanoes by themselves.[2] New Guinea's active volcanoes are mostly coastal, often forming offshore islands that trace arcuate lines of crustal-plate convergence. Rarely, it seems, is one or other of them not active, but it was a particularly active period when William Dampier arrived there in 1700, encountering five of these 'burning' islands. All night on 25 March, he reported that the first

> ... *vomited Fire and Smoak very amazingly; and at every Belch we heard a dreadful noise like Thunder, and saw a Flame of Fire after it, the most terrifying [sight] that I ever saw.*[3]

What of New Guinea's southern neighbour, Australia, which as we have seen is today a somewhat geologically passive place? Have there been volcanic eruptions there within the 65,000 years or so that people have lived there?

Not surprisingly when you consider the vastness of Australia, the answer is yes. And yes, too, there are extant stories about some of these eruptions, which also comes as no surprise given the likelihood that Australia's Indigenous peoples have preserved other stories for perhaps more than 10 millennia. Yet critically, these eruptions have been isolated in both time and space in Australia, making any associated oral traditions relatively easy to link to particular events and places. This condition makes Australia an excellent place to study the antiquity of these oral traditions, better for this purpose than New Guinea, Japan, Italy or Iceland, where the ubiquity of eruptions through time has rendered the relevant bodies of oral tradition far more nebulous and difficult to interpret.

The map in Figure 6.1 shows places in Australia where volcanic eruptions are known to have occurred within the past 65,000 years or so. They include at least five sites in northern Queensland (Lakes Barrine, Eacham and Euramoo; Kinrara and Nulla), five in southern Victoria (Napier, Eccles, Tower Hill, Warrnambool and Leura) and three just across the border in South Australia (Muirhead, Schank and Gambier). The activity of the volcanoes that these features represent must all have been witnessed by Aboriginal peoples and, as we shall see below, some of their observations have come down to us today.

It is something of a puzzle that young volcanic activity should have occurred on an ancient continent. Indeed, scientists are still struggling to explain some of this activity, although that in the south is possibly associated with the presence of a massive body of liquid rock – a mantle plume – that today lies beneath much of Tasmania and the adjacent ocean floor. This plume is a 'hotspot', a place where liquid rock from the Earth's mantle protrudes into the overlying solid crust, coming so close to its surface (perhaps 80–150km/50–95 miles below it here) that some of this liquid rock has occasionally escaped upwards along cracks and come out at the Earth's surface. It is not beyond the bounds of possibility that this might also occur in the future.

We know this mantle plume has been in existence for several tens of millions of years, for it first encountered the Australian mainland more than 33 million years ago to form the Mt Jukes Volcano (Cape Hillsborough), near what is now the central eastern Queensland coast (see Figure 6.1). As the Indo-Australian Plate continued its slow, inexorable progress north-northeastwards, as described in Chapter 2, so the mantle plume passed under the continent and formed a line of volcanoes, just like the classic hotspot chains of oceanic islands.[4] The resulting Cosgrove Hotspot Track is some

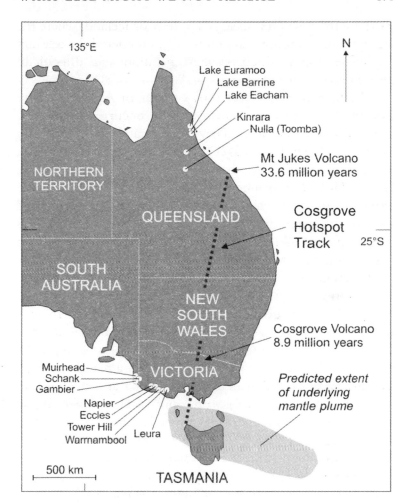

Figure 6.1 *The eastern half of Australia showing all the volcanoes known to have erupted within the last 65,000 years, the probable time that Aboriginal people have occupied the continent. Four of these volcanoes are shown in northern Queensland, five in southern Victoria and three in the nearby part of South Australia. The origin of at least some of this comparatively young volcanic activity can be attributed to the presence of the mantle plume that created the Cosgrove Hotspot Track within the last 33 million years.*

2,200km (1,367 miles) in length, the longest hotspot track yet to be documented on a continent.

This mantle plume is implicated in the young volcanic activity that occurred in southern Australia, but exactly how is still something of a mystery. The volcanic activity defines the Newer Volcanics Province (NVP) and includes at least 704 points of eruption from at least 416 separate volcanic centres within an area of some 19,000km^2 (7,336mi^2). But as you can see from the map, the NVP does not lie on the Cosgrove Hotspot Track – it is not even really close. Yet, given that the earliest volcanic activity in the NVP began about five million years ago, the timing with the progress of the relative movement of the plume is about right; it would have been closest to the NVP at this time. So we can envisage a situation where a sizeable piece of the plume (a diapir) at this time branched or broke off to its west to form a body of liquid rock beneath this part of southern Australia where, for the next five million years or so, it has periodically leaked to create the NVP.[5]

Uncertainty feeds scepticism, especially when this fuels self-interest. The idea that scientists cannot explain why volcanic activity has occurred recently in Australia might indeed strain the credulity of some of its citizens about whether this activity really did occur. But we have two excellent ways of corroborating this: radiometric dating and, more tellingly for sceptics, eyewitness accounts.

Largely because its results contradict certain narrow partisan interpretations of the Christian Bible, the concept of radiometric dating has received a bit of bad press in some quarters,[6] although it has been used routinely for decades by scientists to determine the ages of particular rocks. After these rocks first form, radioactive impurities within them start to change – or decay – over long time periods into different types of material. By knowing the rates at which such radioactive decay takes place, we can calculate the time that has elapsed since the formation of a particular rock by measuring the proportions in it of both the 'parent' radioactive substance and the 'daughter' (or decay product). Most volcanic rocks that have been dated, including those from Australia's younger volcanoes, have been subject to age determination using the potassium–argon (K–Ar) method. In this, an isotope of potassium (^{40}K) decays to one of argon (^{40}Ar). For half the ^{40}K present in the original rock to become ^{40}Ar, it takes around 1.3 billion years, so you can see potentially how powerful this dating technique is. Almost every volcanic rock that became solid (cooled from a liquid state) within the past two billion years or so can be dated using this method.[7]

Another important radiometric dating technique is radiocarbon dating, based on the decay of the radioactive isotope of carbon (^{14}C), and applicable to any substance with carbon in it. By measuring the ratio of ^{14}C to the non-radioactive carbon isotope ^{12}C, the age of a substance like a piece of shell, bone or hair can be measured. Its age is that from the time when the carbon-containing lifeform died, ceasing to renew its stock of ^{14}C from the food it ate, until the present. With a half-life of 5,730 years and with quantities of the two isotopes able to be measured very precisely, radiocarbon dating has been used widely. You cannot use it on most volcanic rocks, but you can use it on some of the things that are buried when lavas flow (or hot rocks are blown) out of volcanoes. They include items like soils or shell middens, and even old fire pits containing charcoal and bones. Some of Australia's young volcanoes cannot be dated directly because the erupted materials are too weathered; in such cases, radiocarbon dates from associated materials can be used to provide maximum (sometimes minimum) ages for a particular eruption.

There was once a man, a veritable giant named Craitbul, who with his wife, also of 'immense size', and two sons wandered through south Australia in search of a place where they might live in peace, free from fear of the evil spirit Tennateona.[8] The family first settled atop Mt Muirhead, one of the younger volcanoes in the NVP, where they lived 'for a considerable time in peace'. Accustomed to cooking food in the ground, they dug a hole there in which they placed their food, covered it with earth and went to sleep, anticipating that it would be cooked when they got up the next morning.

But in the night they were awoken suddenly by the shrieking of a bird known as a *bullin*. Believing this portended the presence nearby of Tennateona, they fled, eventually settling on Mt Schank, where once again they camped and 'began to enjoy themselves'. But one night, their cooking oven emptied, they were again awoken by the *bullin*'s cries and fled inland, away from the sea to which they believed Tennateona preferred to be close. They travelled to Mt Gambier 'and lived here a long time', as usual making their oven within it. But one day water rose up from below, putting out their oven's fire. So they dug another, and the same thing happened, and so on and so on. Finally they gave up and were last heard of living in a 'cave on the side of the peak'.[9]

If we equate the ovens with the craters within these mountains, then we could also read the cries of the *bullin* and the emptying of ovens as euphemisms for the alarm accompanying the eruptions of these volcanoes, as well as the facts of this. The four craters of Mt Gambier are *maars*, products of spectacularly explosive eruptions that occur when superheated liquid rock (magma) rising up a vertical fissure through the crust encounters cold groundwater. The resulting phreatomagmatic explosion shoots out lava fragments and ash in every direction. When these fall to the ground they often harden to form a circular ridge – marking the spatial limits of the eruption – which may later become filled with rainwater to form a *maar* lake. Most of the *maars* of the Mt Gambier region are today water filled, so we can interpret Craitbul's ovens as being the active *maar* volcanoes that were subsequently flooded. It is hard to see what other interpretation might be placed on such a detailed story (see colour plate section).

How old might the story of Craitbul be? Radiocarbon dating of plant material buried within volcanic ash on the floors of the *maar* lakes of Mt Gambier suggests that the most recent period of volcanic activity here took place around 4,300 years ago, about the same time as the last eruption of Mt Schank.[10] Several other Aboriginal stories report fire being thrown from these mountains, and at least one informant at Mt Leura identified some of the volcanic bombs, which are typical products of explosive eruptions, as just like 'stones which their forefathers told them had been thrown out of the hill by the action of fire'.[11]

Mt Eccles is another young NVP volcano,[12] and there are said to be Gunditjmara Aboriginal stories that recall its most recent eruption. This is likely to be that of the Tyrendarra lava flow, much of which is now as much as 33m (108ft) below the ocean surface. Yet when it formed it covered only dry land; it did not run into the sea. So while the lava flow itself has proved impossible to date directly, we can estimate a minimum age of 10,500 years ago for the eruption that formed it – the time when the sea floor now 33m deep was last dry land.[13]

Much of the realisation that Aboriginal Australians had witnessed volcanoes erupting came from understanding the meanings of place names. Among some of the early European settlers in parts of Australia, there was a tradition of recording Indigenous names for particular places, something that has occasionally been found to hold insights into their history.

There never were elephants living in Australia, so the name Mt Elephant for the young scoria cone volcano in south-west Victoria is clearly one given by European settlers, one assumes on account of the pachydermic appearance of the mountain from a certain angle. Yet far earlier in time, the Aboriginal people of the area called this mountain Gerinyelam, meaning 'hill of fire'. Given that Mt Elephant formed 5,000–20,000 years ago, we can assume that its eruption was witnessed, a fact that was probably once preserved in stories that now appear to be lost, yet have their essence kept alive in the name Gerinyelam.[14] Another example is that of the Gunditjmara people who call nearby Mt Eccles by the name Budj Bim (high head), and its associated spatter cones and tumuli *tung att*, or 'teeth belonging to it'.[15]

What of the other group of youthful volcanoes in Australia, those in northern Queensland? Their origins are perhaps even more elusive than those of the NVP.[16] Like this group of volcanoes, they exhibit no age progression in any direction, so that eruptions here cannot be readily explained by the movement of the Australian continent over a mantle plume, as can eruptions of the older volcanoes along the Cosgrove Track. Eruptions in northern Queensland have been continuing – on and off every 10,000 years or so – for the last nine million years, and the area is still considered 'potentially active'.[17]

Aboriginal stories about eruptions in the Kinrara area belong to the Gugu Badhun people and describe events that may have occurred 7,000 years ago.[18] One story recalls the time when the watercourses that flow through the area – and from which the local Aboriginal people rarely strayed far – caught fire. Another story is about a witch doctor who made a huge pit in the ground, filling the air with dust, in which people became disoriented and died. This eruption of Kinrara involved the emplacement of lava flows – some ran down the river valleys for tens of kilometres – which is a plausible explanation for rivers said to have caught fire. The dust cloud and the deaths of affected people are equally easy to explain, for this eruption – the youngest known at Kinrara – was focused on a crater (the witch doctor's pit) and involved ash-rich eruptions likely to have created dense clouds suffused with poisonous gases that drifted over the surrounding area, killing many of its inhabitants.[19]

Further north, Lakes Barrine, Eacham and Euramoo on the Atherton Tableland are three *maar* crater volcanoes, one or more of which may have erupted during the time that people have been living in this area. By dating (using radiocarbon) the earliest layers of sedimentary fill in these craters, minimum ages for their formation – presumably during a volcanic eruption – have been obtained. In this way, we know that the most recent possible date for the last eruption at Lake Eacham was around 9,130 years ago; that at Lake Euramoo perhaps 15,000 years ago; and that at Lake Barrine some 17,300 years ago.

Several Aboriginal groups in the area have stories about the eruption of Lake Eacham (and Barrine and Euramoo in some versions), all of which are quite similar. The essence of these stories is that two newly initiated youths were confined to camp, as was common practice, while their elders went to collect food. Needing to defecate, one youth went a short distance into the bush, accompanied by the other – breaking the taboo. The boys saw a wallaby and one fetched a spear that his companion threw at the animal. It missed, instead hitting a flame tree that is sacred to the Rainbow Serpent, the most iconic supernatural being in the Dreamtime bestiary. Thereafter things began to go badly wrong, and when the elders returned from hunting, they knew it. For although it was the middle of the day, the sky turned blackish-red – the colour of a striking sunset – and shortly afterwards the earth began to 'crack and heave, spilling out a liquid that engulfed the camp and all the people in it, and forming the three modern crater lakes'.[20]

Australian Aboriginal stories about volcanic eruptions have come to confirm what science has learnt only far more recently. We can rate the authenticity of these stories highly, for nowhere do we find them applied to features that are not youthful volcanoes. In northern Queensland, for instance, Aboriginal stories about eruptions do not refer to other, much older crater lakes like Lynch's Crater, where the last eruption took place around 200,000 years ago, well before people first reached Australia. And in the NVP in South Australia, no Aboriginal eruption stories are applied to those volcanoes – like Mt Napier and Tower Hill – where the last eruptions occurred more than 30,000 years ago.

There are few other places in the world where indigenous stories, passed on orally, recall ancient volcanic eruptions that have been virtually unknown in modern (literate) times. Of course, such stories abound where there are currently active volcanoes that have been periodically active ever since people arrived in their vicinity. So while any associated oral traditions are of interest, they do not really hold any surprises.

Take the iconic Mt Fuji (Fujisan), a mere 100km (62 miles) from Tokyo on Honshu Island in Japan. Aside from its form – that of a classic volcano, which has been celebrated in Japanese art and spirituality for millennia, Fujisan has been periodically active. Although its last proper eruption was in AD 1707–1708, the mountain groans audibly at times and belches gases continuously from fumaroles along its periphery, all signs that this volcano is far from deceased. There are numerous oral traditions about Fujisan's origin, one being about a giant who decided one day to fill up the Pacific Ocean with earth. Toiling all night, he was dismayed when dawn broke to find that he had made hardly any progress with his task, so he dumped the bag of earth he was carrying where Fujisan is today … and stomped off sulkily. While entertaining and possibly containing some geological detail – perhaps about how Fujisan was created quickly – the story does not supply any details that are not already glaringly obvious. Similar stories abound in Icelandic lore, for example, as well as elsewhere.[21]

In the islands of Hawaii there are numerous stories about volcanic eruptions, but here – with a degree of precision that is almost unique – it is chiefly genealogies passed on orally for several hundred years that can be used to date particular events, something that science has since corroborated. For example, Hawaiian traditions (*mo'olelo*) recall that during the reign of Keali'iokaloa (approximately AD 1525–1600), a young chief from the Puna district of the Big Island (often spelled Hawai'i to distinguish it from the name of the island group) failed to properly acknowledge Pele, the fractious volcano goddess, so she pursued him downhill at high speed and followed him into the ocean – a likely recollection of a lava eruption dated to the same period.

There are three other examples of stories about volcanic eruptions that are worth mentioning since they contain significant information about particular events that science has not yet been able to fully verify.

The first example comes from the north-west United States, where lies Mt Rainier, perhaps the most potentially hazardous volcano in the Cascades. Rising almost 4,400m (14,436ft) above the ocean, Rainier can be clearly seen from most of Seattle and other densely populated coastal areas of Washington State. Rainier has been periodically active for most of the 15,000 years or so that people have lived in North America, and they have witnessed – and presumably struggled to rationalise – the multifarious expressions of its activity. In a Nisqually Native American oral tradition, Rainier is a monster that moved through the landscape swallowing every creature it encountered, until one day it met a fox that had wilily fastened itself to a neighbouring mountain. Exerting itself to try and devour the fox, Rainier eventually burst a blood vessel, causing rivers of blood to flow down its flanks. It is suggested that this myth recalls the time some 5,600 years ago when a phreatomagmatic eruption at the top of Mt Rainier produced the phenomenal Osceola Mudflow, which travelled more than 120km (75 miles) (to the outskirts of modern Seattle), eventually covering an area of more than 200km² (77mi²). Studies of the nature of the flow structures within the mudflow sediments show that it moved as much as 19m (62ft) a second across the Puget Sound lowland – by any yardstick, a dramatic and memorable event worthy of becoming the subject of an important and enduring myth.

The second example comes from Indonesia, some of the islands of which are spectacularly volcanically active, lying above the place where the Indo-Australian Plate is being thrust beneath its Asian counterpart. The 1883 eruption that destroyed the island of Krakatau (Krakatoa) and was heard across 8 per cent of the Earth's surface occurred here for this reason.[22] Plate underthrusting and associated melting at depth also explain the brooding presence of the volcano of Tangkuban Parahu in western Java, 17km (11 miles) north of the city of Bandung, home to some three million souls, most of whom have witnessed (or will witness in their lifetimes) activity at this volcano (see colour plate section). A Sundanese oral tradition, first referred to in the *Bujangga Manik* (56 lines of narrative verse written about 500 years ago on 29 palm leaves), explains that a young man named Sangkuriang was once whacked over the head with a *centong* (rice scoop) by his mother, Dayang Sumbi, sustaining a wound that left a permanent scar. Sangkuriang and his mother parted company after this, but when he later went in search of a wife he fell in love with her, not knowing that she was his mother, and determined to marry her. But Dayang Sumbi felt the scar on his head and knew he was her son, so refused to marry him. Sangkuriang was undeterred, so Dayang Sumbi challenged him to do what she thought was impossible – to create in one night a huge lake and a giant boat (*parahu*) for them to sail there. Damming the Citarum River and felling a giant tree, Sangkuriang almost did the impossible. Seeing this was about to happen, Dayang Sumbi caused the dawn to rise early that day and the cocks to crow. Believing then that he had failed, Sangkuriang smashed the boat, turning it upside down (*tangkuban*), and eventually both he and his mother drowned when the dam broke and the lake emptied. It is the existence of a former lake atop Tangkuban Parahu that gives some authenticity to this story as deriving from an eyewitness account. While poorly dated, it is clear that the Citarum Valley was once dammed by landslides, causing a lake to form and eventually to empty, perhaps 16,000 years ago.[23]

A final example comes from Eurasia, from the breathtakingly beautiful Caucasus Mountains that mark the border between Georgia and Russia, and which formed as a result of the compression between the Arabian and Eurasian crustal plates. Numerous volcanoes, the highest of which are snow covered, tower over the landscape. Many have been active within the last few thousand years, a clear sign that a mantle plume has formed beneath the mountain range, exploiting lines of crustal weakness created during the prolonged compression. Even a passing acquaintance with Caucasian (*sensu stricto*) cultures is likely to throw up the name of Amirani, a hero of the region best remembered for having been chained to one of its mountains and attacked – like the Greek Prometheus and the situation at Mt St Helens (see colour plate section) – by an eagle drawing his blood. Another common story about Amirani relates how he cornered hordes of demons within the walls of a mountaintop fortress, then proceeded to massacre them; so much blood was spilt that the fortress walls finally gave way and rivers of blood poured down the sides of the mountain.[24] Like the Promethean stories about Amirani, this one seems likely to recall a volcanic eruption, but from where and when? Within the Caucasus, the greatest concentration of Amirani traditions appears to be associated with Mt Kazbek volcano, which has not erupted for about 6,000 years. A better candidate for the Amirani stories may be Mt Elbrus, the highest in the range, which last erupted about AD 320 and may be due for another period of activity soon.[25]

A summary of the likely longevity of the examples of volcano stories, preserved for most their time in oral form, is shown in Figure 6.2 overleaf.

There is no question that ancient stories of volcanic eruptions and kindred phenomena exist in many of the world's cultures and may extend back many millennia, perhaps 10 millennia in the cases of Eccles and Eacham, and possibly even more for Tangkuban Parahu. Science is indebted to indigenous stories of this kind, not simply because they confirm the existence – and sometimes give insights into the nature – of particular volcanic events, but also because in some cases they identify events that may not hitherto have been detected by scientific investigations.

The upper parts of our oceans are alive, yet much of what lives there is so small that we cannot see it with the naked eye – so often we naively assume the ocean water we swim in to be clean. It is not. Microscopic carbon-based organisms, often lumped under the name plankton, abound. When they die, their hard parts usually sink to the ocean floor, sometimes kilometres deep, where they may become buried by layers of sediment. They are out of sight but certainly not out of mind for heterotrophic bacteria,[26] which eventually find and consume them, releasing methane gas in the process. Pressure forces this gas upwards from the warm places within the sediment pile to its surface – the ocean floor – where it is much colder. Sometimes ocean water here becomes frozen around handfuls of gas, forming layers of methane hydrates that can themselves become buried at shallow levels within ocean-floor sediment. Now imagine this happening when the sea level was lower, as it was by

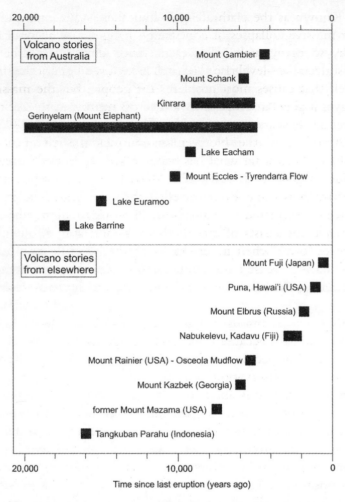

Figure 6.2 *Probable ages of volcanic eruptions recorded in oral traditions from Australia and elsewhere, details of which are in the text.*

120m (400ft) or so during the last ice age; then what happens subsequently as the sea level rises rapidly thereafter. Suddenly the weight of seawater above the layers of methane hydrates – layers of ice-encaged lumps of methane – increases, perhaps to a point where the ice cages are smashed and the methane escapes. The whole pile of sediments in which the gas hydrates had existed becomes buoyant, and is destabilised and liable to collapse.

Known as the clathrate–gun hypothesis, this explanation for massive collapses of continental fringes may account for why so many such collapses occurred during the time of postglacial sea-level rise.[27] In such cases it is often not the slide itself that causes most problems for people, but the massive waves it generates that can wash across nearby coasts. A little over 8,000 years ago, a sizeable chunk – more than 3,000km³ (720mi³) – of the Norwegian continental shelf abruptly collapsed, creating what has become known as the Storegga Slide. It generated waves 12m (40ft) high that swept across nearby coasts and even travelled further afield. The main wave is estimated to have been 20m (65ft) high when it reached the coasts of the Shetland Islands, and a couple of metres lower when it smashed into the Scottish mainland, inundating coastal settlements with little warning and scattering the stone tools used by their inhabitants across a wide area.[28]

No oral traditions of this event have come down to us today, perhaps because it was so long ago and the subsequent admixing of people in many of the affected areas militated against the preservation of ancient stories. You can bet there *were* oral traditions about the Storegga Slide in Norwegian and Caledonian cultures, reflected echoes of which may be found in their extant myths about the destruction of worlds,[29] although no specific connections have been identified. The same is not true of other places where land has abruptly disappeared.

In island worlds, it is possible for entire islands to disappear comparatively quickly,[30] something that would clearly have caused consternation among the people of neighbouring islands, who often encoded their observations of such events in myths. To understand how an island might disappear, consider that most islands – especially those of volcanic origin in the middle of oceans – are both steep-sided and may have only a few per cent of their total mass exposed above the ocean surface. Just as happens on any steep-sided landform, landslides occur regularly on the sides of these islands, particularly those that are occasionally rocked by earthquakes.

Such landslides – or flank collapses as geologists more commonly refer to them – may take a chunk out of the side of an island or even remove its top, its entire above-sea portion. In the greater scheme of volcano and island evolution, this is not really significant, but to the air-breathing occupants of the island and its neighbours it is massive.

One comparatively well-documented example of such a 'vanished island' comes from the Solomon Islands archipelago in the south-west Pacific Ocean, where once an island named Teonimenu existed. Even though it disappeared hundreds of years ago, people living on nearby Ulawa Island know where Teonimenu was. Today it is a favoured fishing ground of theirs, with the top of the ancient island now being a submarine bank (named Lark Shoal) some 10m (33ft) below the ocean surface. The Ulawans say that on a placid sea, watchers may look down from boats above and see the ancient structures that people once made on Teonimenu, and sometimes even hear the beating of their drums and the sounds of their voices ...

Like Ulawa today, Teonimenu was once an emergent part of an undersea ridge, steepest on its eastern side where the sea floor descends to the Cape Johnston Trench, where a crustal plate further east is (or was) being thrust beneath that to the west. The area is often shaken by earthquakes, perhaps caused by nearby underthrusting of one plate beneath the other or – more likely given the origins of most of the area's recent seismicity – the north–south compression of the Earth's crust here. For the Solomon Islands rise from an isolated chunk of the Earth's crust that is being compressed obliquely between the northward movement of the Indo-Australian Plate and the north-westward movement of the Pacific Plate. The result is that the Solomon Islands are uncommonly subject to earthquakes, some of considerable magnitude, which often generate tsunamis and may cause land to rise or fall abruptly. It is likely that, one day a few hundred years ago, a sea-floor earthquake close to Teonimenu caused tsunamis but also shook the island so much that it slipped a few tens of metres downslope in a massive flank collapse, removing itself in an instant from the human geography of the central eastern Solomon Islands.

The story has reached us today through the mouths of local people, who will tell you about mad jealous Roraimenu from nearby Ali'ite Island, who married a capricious black-eyed beauty from Teonimenu named Sauwete'au. Tiring of Roraimenu, Sauwete'au one day eloped and returned to Teonimenu. Without compromise in mind, her husband went to consult a spell-seller on a nearby island, then sailed for Teonimenu prepared to plant the wave curse he had purchased there in order to destroy his errant wife's island. As he sailed on his way there, grim-faced, people on nearby islands, alerted to his purpose, called out to him in fear, 'Which island are you going to destroy?' (*Hanua i hei oto a nai warea?*), but Roraimenu did not respond. Placing the curse on Teonimenu, he then retreated to a high point on Ali'ite to watch the destruction of Sauwete'au's island. The water around Teonimenu churned and it was hit by a succession of giant waves as it sank beneath the ocean surface, never again to be seen. Oral histories from neighbouring islands recall how some inhabitants of Teonimenu saved themselves and where their descendants live today. Clearly the disappearance of Teonimenu resonated throughout this part of the Solomon Islands.[31]

For a second example of an island collapse, we turn to the Marquesas Islands of French Polynesia in the Central Pacific. The Marquesas are volcanic and are something of an enigma to oceanic geologists, for the total volume of material making up the islands today is just a fraction of that which has been eroded off them and encircles them in a vast debris apron. The inescapable conclusion is that these islands have many times built themselves up to a state rather like they are in today, before being almost wholly eroded, then rebuilding themselves, and so on and so on.[32]

There is today a rather insignificant uninhabited rock named Fatu Huku in the Marquesas, but when Captain James Cook visited these islands in 1774 it was much larger and probably populated. Yet Porter's map of 1820 showed Fatu Huku to be much as it is today, perhaps a twentieth of its original size, and thus likely to have been affected by a collapse that led to the disappearance of most of the island within the interregnum.[33] Believing that Fatu Huku was once supported by a pillar of rock rising from the Earth's interior (it was not), Marquesan legends attribute its collapse to the shark guardian of the island which one day – to punish its dissolute inhabitants – used its tail to lash the pillar, causing it to break and the island to topple over into the sea.

Something similar is found in Native American traditions, especially along the coasts of Cascadia where massive earthquakes, some of which lowered coastal lands abruptly, periodically occur. Consider the Yurok story in which

> … *Earthquake and Thunder were wandering across the land, discussing the form it should take. When they decided that the ocean should come farther inland, Earthquake told his companion, 'It will be easy for me to do that, to sink this prairie'. So as they traveled [sic] together, 'They kept sinking the ground. The earth would quake and quake again and quake again. And the water was flowing all over'.*[34]

Yet in this region, as elsewhere, observers tended to place more emphasis on the earthquake-generated waves – the tsunamis – than the land movements that actually cause the disappearance of coastal land.[35]

This brings us to Atlantis, the fabulous island-continent described by Plato around 360 BC, which he wrote had abruptly and cataclysmically vanished beneath the waves. Atlantis never existed and no amount of wishful thinking – and there is certainly an abundance of that – can make it otherwise, but Plato wanted to make his story sound as realistic as possible, so he stitched together details from actual events to create his enduring fiction. Since he required Atlantis to disappear, it follows that many of these details referred to land sinking, often near instantaneously, examples of which were well known to scholars in Ancient Greece.

Many earthquakes occur on the Greek peninsula and in the Aegean Sea. The Gulf of Corinth, west of Athens, has been a major focus of earthquakes in recent times, as it undoubtedly was during the time of Plato. We have some information about ancient earthquakes there. For example, in 464 BC, a mere 37 years before Plato was born, an earthquake flattened the city of Sparta and was probably much talked about during Plato's formative years.[36] For earlier evidence of the power of earthquakes to destroy cities, Plato need have looked no further than the island of Crete, where the Minoan settlements (their ruins still visible) of Agia Triada and Phaistos had been destroyed by massive earthquakes about 1700 BC.

Then in 426 BC, when Plato was just a baby, the Maliakos Gulf Tsunami affected parts of Greece, and it is likely that the reports of this came to influence Plato decades later when he wrote his story of Atlantis. At the time of this memorable tsunami, the Peloponnesian War between Sparta and Athens was in its sixth year and would last another 14. In this particular year, according to Thucydides, the principal chronicler of the war, the Spartans (heading the Peloponnesian League) were ready to invade Athenian territory,

> ... but numerous earthquakes occurring, turned back again without the invasion taking place. About the same time that these earthquakes were so common, the sea at Orobiae, in Euboea, retiring from the then line of coast, returned in a huge wave and invaded a great part of the town, and retreated leaving some of it still under water; so that what was once land is now sea; such of the inhabitants perishing as could not run up to the higher ground in time. A similar inundation also occurred at Atalanta, the island off the Opuntian Locrian coast, carrying away part of the Athenian fort and wrecking one of two ships which were drawn up on the beach.[37]

While it is likely that some of this detail was written down in Plato's time, Classical Greek culture was still largely an oral one. Plato would routinely have heard the tales of travellers from distant lands, and his imagination would probably have been stirred by some of the outrageously fanciful explanations for natural phenomena they related, his mind likely wondering at the gullibility of people and plotting plausible scenarios for Atlantis, his ideal society gone bad.[38]

On 30 June 1908, a sizeable rock – probably an asteroid some 60–70m (200–230ft) in diameter – entered the Earth's atmosphere. Between 6 and 10km (4–6 miles) above the Earth's surface in Tunguska, Siberia, it exploded. The blast was powerful enough to flatten more than 2,000km^2 (772mi^2) of the larch forest that swathes the natural *taiga* landscape in this sparsely populated part of modern Russia.[39] When scientists finally surveyed the area almost 20 years later, they had to be 'indirect and circumspect' when questioning Evenki eyewitnesses to this event, since many believed it had been a visit by their fire god, discussion of which – as for the Klamath entering the vicinity of Crater Lake in Oregon (see Chapter 1) – was awkward because of its religious and cultural connotations.

Asteroids, comets and meteorites (collectively known as bolides) regularly enter our planet's atmosphere, almost always being fragmented above the ground surface but more often than not – unlike the situation at Tunguska – smashing into the Earth's surface. Most bolides fall into the ocean, but some of course hit the land; impact craters pepper the landscape of every continent.

Bolide impacts are undoubted memorable events, worthy subjects for oral traditions, particularly in places where the extraterrestrial object hit the ground and created an impact crater. In Chapter 1 we saw how a bolide impact more than a millennium ago in Italy fast-tracked the people's adoption of Christianity. Features like impact craters often remain visible in landscapes for hundreds of thousands of years, a constant reminder to local people of the events that created them. In the same way that some cultural groups venerate gods within active volcanoes, beseeching them not to direct eruptions across their fields and villages, so too are there records of group worship at impact craters. Consider this translated description from the 1920s of local traditions at the Campo del Cielo (Field of the Sky) site in the Argentine pampas, where there are numerous impact craters as well as scattered fragments of a giant iron meteorite:

> The meteorite of the Chaco [province] was known since earliest American antiquity through stories from the Indians who inhabited the provinces of Tucumán. [These people] had trails and easily traversed roadways that departed from certain points more than 50 leagues [more than 278km/173 miles] away, converging on the location of the bolide. The indigenous tribes of the district gathered here … in veneration to the God of the Sun, personifying their god in this mysterious mass of iron, which they believed issued forth from the magnificent star. And there, in the stories of the different tribes of their battles, passions and sacrifices, was born a beautiful, fantastic legend of the transfiguration of the meteorite on a certain day of the year into a marvellous tree, flaming up at the first rays of the sun with brilliant radiant lights and noises like one hundred bells, filling the air, the fields and the woods with metallic sounds and resonant melodies to which, before the magnificent splendour of the tree, all nature bows in reverence and adoration of the Sun.[40]

This story may recall the arrival of this large meteorite, how it 'flamed' brighter than the sun, and the sounds that were heard as it exploded into pieces and smashed into the ground in dozens of different places within a short period of time.

To enable an understanding of how non-literate peoples may have recorded for posterity their observations of bolide falls and impacts, Australia again provides some useful examples. Not only is its Indigenous culture one of considerable unbroken longevity and demonstrably effective in preserving ancient traditions, but Australia – being comparatively dry in most parts – has environments that are particularly well suited to the preservation and study of bolide impact structures. The rocks – known as tektites – that form when a meteorite impacts the ground surface are strewn around the impact site (in what geologists call strewnfields), but are often found much further away. So powerful are some impacts that in an instant they cause the earth-surface rocks to melt or vaporise, creating small pieces of molten material that can be widely dispersed. Some even enter the atmosphere and are carried well away from the impact site; for example, Australian tektites are found in China and Antarctica as much as 11,000km (6,835 miles) from their impact site.[41]

There are 27 known bolide impact craters in Australia, most in the (drier) Northern Territory and Western Australia. It is something of an enigma that there are comparatively few Aboriginal stories about the formation of these craters, yet comparatively many describing meteorite falls. Hints that local Aboriginal people knew the origin of the Henbury meteorite craters in the central Australian desert come from the story that told that they should not camp near the place, which they called *chindu chinna waru chingi yabu* ('sun walk fire devil rock').[42] The Henbury meteorite falls took place around 4,200 years ago.

But understanding is not necessarily witnessing, and there are Australian Aboriginal stories about the origins of meteorite craters that formed long before people first reached Australia (see colour plate section). In these situations, it is obvious that the details of eyewitness accounts of particular events were transferred to landforms that were patently of the same origin, despite being far older.[43]

In South Australia on the edge of the Nullarbor Plains, swathes of which are littered with meteorite fragments, there is a typical meteorite impact story. Belonging to the Adnyamathanha (Wirangu) people, a retelling for the school curriculum of the original story explains that:

A long, long time ago, a huge meteorite hurtled towards the earth from the northward sky, and smashed into the ground near Eucla. Because it was so big, a dent appeared in the crust of the earth and the meteorite bounced high into the air and out into the Great Australian Bight [ocean] where it landed with an enormous sizzling splash. It was hot from its trip through space so it gave off a good deal of steam and gas as it sank through the waves. But this was no ordinary meteorite. In fact, it was

*the spirit Tjugud. In the deep water near by, the spirit woman
Tjuguda lay asleep. All the noise around her woke her up and
she was very angry. She bellowed and the elements roared with
her. The wind blew, the rain pelted from the sky and the dust
swirled.*[44]

Given that Aboriginal eyewitness accounts of meteorite falls
in Australia are more common than impact craters, it is likely
that – as at Tunguska – many of these events were airbursts, a
solid bolide shattering above the ground surface and producing
a rain of fragments, none of which proved large enough to
create a visible crater when it hit the Earth's surface. One
example from New South Wales (a state without a single
impact crater) recalls that:

*The sky heaved and billowed … the stars tumbled and clattered
and fell one against the other. The great star-groups were
scattered, and many of them, loosened from their holds, came
flashing to the earth. They were heralded by a huge mass, red
and glowing, that added to the number of falling stars by bursting
with a deafening roar and scattering in a million pieces which
were molten … The disturbance continued all night. When the
smoke and clamour had died away and morning had dawned
it was seen that the holes had been burnt into the earth, and
great mounds were formed by the molten pieces, and many caves
were made.*[45]

We cannot know the antiquity of this story because no
airburst event has been identified in the area, but given the
abundance of similar stories in Australian Aboriginal cultures,
there is little reason to doubt that it is an authentic memory
of a real event.

In other parts of the world, preliterate descriptions of
meteorite falls are likewise somewhat vague and difficult to
tie to any single event, something that would allow us to
determine the antiquity of these stories. Just as there are
stories in many cultures of a global flood, which are likely to
be a palimpsest of stories of successive floods in particular

places,[46] so there are stories of a world fire that wiped out humanity in some cultures. In some of the associated stories, a cosmic origin is identified for the fire, suggesting that this tradition derived from successions of meteor showers or airbursts in particular locations.

Finally in this chapter, we turn to a little-studied topic of oral traditions, namely their potential for recollecting the existence of animals that are now extinct. The real challenge here is to move beyond the explosion in imagined creatures that has characterised post-nineteenth-century fiction and today transfixes many audiences, often confusing people's ability to distinguish fantasy from past realities.[47]

One of the most extraordinary stories of this kind concerns one of our kin, *Homo floresiensis*, which is known to have co-existed with our own species on Flores Island in Indonesia until perhaps 12,000 years ago.[48] The discovery was made in 2003 in a limestone cave named Liang Bua, 25km (16 miles) from the sea. At a depth of nearly 6m (20ft) in sediment layers dating from about 18,000 years ago, the remains of this small-bodied, chinless bipedal hominin were found. The young age for the remains of *Homo floresiensis* puzzled scientists for a long time, but a recently revised age for the holotype at Liang Bua to some time between 60,000 and 100,000 years ago sits more comfortably with most.[49] This is because our species, *Homo sapiens*, was pretty intolerant of other hominid species, and adept at eradicating them shortly after coming into contact with them. The stone tools found at Liang Bua suggest that the *Homo floresiensis* people living there continued doing so until some 50,000 years ago, which fits with the revised age for the hominid bones, but there are hints from oral traditions that they may have survived into more recent times.

Long before the discovery of *Homo floresiensis* remains, the people of Flores Island had stories about the *ebu gogo*, the name given to 'short, hairy and coarse-featured' hominoids said to have lived on the island until quite recently. In 1984, when anthropological researchers started work in the Nage region of central Flores, they were told a story by descendants of the inhabitants of the (former) village of 'Ua how its people had once cornered a group of *ebu gogo* in a cave called Lia Ula several generations earlier. Fed up with them abducting children and stealing their crops, the people of 'Ua set fire to a quantity of palm fibre and immolated these *ebu gogo*.[50]

Ebu gogo stories are known from other communities on Flores and may be part of the 'wildmen stories' that are comparatively common in island South-east Asian cultures,[51] and in some other parts of the world. What is of key interest about the *ebu gogo* stories is whether or not they recall the presence until very recently of *Homo floresiensis* on Flores. At present there is not enough information to test this suggestion, especially since science would currently have this hominid species becoming extinct about 50,000 years ago. But it is an intriguing story that science may one day answer more satisfactorily. Another explanation is that the *ebu gogo* stories (not the *ebu gogo* themselves) have endured for thousands of years, ancient cultural memories being transposed onto more recent incidents; perhaps the occupants of the cave at Lia Ula were mere bandits, subsequently labelled *ebu gogo* to help justify their extermination.[52]

We now move to Australia and, again taking advantage of Aboriginal cultural ability to preserve stories millennia old, ask whether there are ancient creatures – perhaps long extinct – that were observed by Aboriginal people who then started the process of commending these stories to posterity. There are indeed such stories. Take that of the *bunyip*, a huge beast said to live, forever groaning and bellowing, in the deepest waterholes, from which it would emerge after dark to prowl the land in search of humans as food. One account from the 1840s reports through a colonist's words the attitudes of Aboriginal Australians towards the *bunyip*:

> ... *a large animal having at one time existed in the large Creeks & Rivers & by many it is said that such animals now exist & several of the Fossil bones which I have at various times shown to them they have ascribed to them. Whether such animals as*

*those to which they refer be yet living is a matter of doubt,
but their fear of them is certainly not the less & their dread of
bathing in the very large waterholes is well known.*[53]

Long-standing Aboriginal beliefs in *bunyip* certainly fed into
the interpretation of sightings of other creatures in recent
times. Consider the observations of cattleman Thomas Hall,
who worked at Canning Downs (New South Wales) and
visited an Aboriginal ceremonial site (a *bora* ring) in about
1858, where he saw a drawing of a *bunyip* painted with 'raddle,
pipe clay and emu oil' on a large tree. Later he saw one, a
creature local people knew as *mochel mochel*:

> *Mr Hall was bringing a mob of brumbies [wild horses] down from
> Swan Creek, and when some distance ahead of his companions,
> was startled by a scream coming from a place known as the
> Gap Creek Junction Hole. Riding over to investigate, he saw
> an animal in shape similar to a low set sheep dog, the colour of
> a platypus, head and whiskers resembling an otter, coming from
> shallow water across a strip of dry land into deep water ... the
> impression made on Mr Hall's mind was that the Mochel Mochel
> or Bunyip was a kind of otter ... [local Aboriginal people], he
> says, had a great dread of it and nothing could persuade them to
> bathe in, or even go near a water-hole believed to be the home of
> the Mochel Mochel.*[54]

In this case, it was indeed perhaps an otter. Yet the extract
exemplifies the point that distant Indigenous memories of the
bunyip – like those of the *ebu gogo* – were perhaps transposed
onto other, comparatively unfamiliar creatures long after
the originals had died out. The avoidance behaviour that
characterised their distant ancestors' attitudes towards *bunyip*
also became transferred by Aboriginal people to these other
creatures, as did their role in Aboriginal ceremonial life.

Most of what are regarded as authentic Aboriginal
descriptions of *bunyip* picture them as 'long-necked, maned,
tusked, [and] horse-tailed', so the discovery of fossils in
numerous Australian contexts of a herbivore, the extinct

marsupial tapir *Palorchestes azael*, 'about the size of a horse ... [and like] no other creature known', has raised the possibility that Aboriginal stories of *bunyip* recall encounters with this animal.[55] Once thought to be a giant kangaroo, *Palorchestes azael* has also been likened to a bull in size and had 'huge koala-like claws, enormously powerful forelimbs, a long ribbon-like tongue and a large elephantine trunk'. It occupied eucalypt woodlands close to shallow lakes or marshes, and once lived across much of eastern and southern Australia. Its possible representation in Aboriginal rock art (see colour plate section) strengthens suggestions that it may be the basis of *bunyip* stories.[56] Direct dating of the fossil remains of *Palorchestes azael* shows that it lived mostly 40,000–50,000 years and more ago, but an age of 23,000 years ago – from surrounding terrace deposits – has been reported for its presence at Riversleigh, now a World Heritage Site, in Queensland. An age for a megafaunal bone cache at Spring Creek in Victoria suggests that this animal may have survived even longer here, perhaps to sometime within the last 20,000 years.[57] Is it possible for Aboriginal traditions of this creature to have survived almost 20 millennia, or does the likelihood of its recollection in those traditions signify that it must have survived until even more recent times?[58]

To put into context Aboriginal stories of the *bunyip* and other possibly legendary creatures that might recall observations of megafauna, consider that the fossil record of Australia shows that it was once populated – like many other continents – by innumerable species of large-bodied animals, or megafauna, most of which are now extinct. The question of why megafaunal extinctions occurred in Australia, as well as in North America, South America and several other regions, has taxed scientists for more than 100 years and still occasionally provokes robust debate. Basically there are two camps: those who believe that, through predation, humans were responsible for rapid megafaunal extinctions; and those who regard climate change, perhaps a rapid cooling event like the Younger Dryas (see Chapter 4), to have led to megafaunal death from starvation and water deprivation. In deciding which explanation is correct or, indeed, whether both contributed in some measure to megafaunal extinctions in particular places, two things are important to consider.

The first is that megafauna are/were big and that, because most smaller animal species clearly did not suffer the same comprehensive degree of extinction at the same time, it follows that their large size made megafauna singularly unable to adapt to whatever it was that caused their extinction. It might be, for instance, that their large size made them especially attractive to bands of hungry human hunters who devised ingenious ways of killing them and, encouraged by their success, continued the blitzkrieg until their prey disappeared, unable to sustainably reproduce itself.

The second consideration is timing. If megafaunal extinctions occurred simultaneously on every continent, then it is a big ask to put the blame solely on people, because then you would need to suppose that they behaved independently in exactly the same ways towards megafauna at exactly the same times – and what are the chances of that? So simultaneous extinctions must point to some global (or near-global) cause, such as rapid cooling or, as in more imaginative scenarios, massive bolide impacts or volcanic eruptions blotting out the sun for several years and plunging the Earth into a prolonged 'volcanic winter'.

The fact of conspicuous megafaunal extinctions in Beringia (a region centred on far-eastern Russia and Alaska), and North and South America, gives us an opportunity to test competing explanations. It is generally assumed, albeit not proven, that the earliest Americans entered the continent through Beringia, thence into North America, and thence into South America. The first humans in the Americas were no fearless explorers determined to push the geographical limits of the known world, but were merely following the trails of mammoths and mastodons, camels and cheetahs, and horses.[59] Radiocarbon ages for extinct megafauna in these three regions do indeed show a time progression suggesting that human overkill was a major factor in these extinctions, but the same is not true elsewhere.

Take Patagonia in the cold far south of South America, for instance. Here megafauna, including bears, horses, giant ground sloths and the various llamas and guanacos (whose smaller surviving relatives are essential beasts of burden throughout this region today) once ranged across the seemingly endless steppes. Humans first arrived in this part of Patagonia just before a cold (stadial) period – the Antarctic Cold Reversal – began. After it ended, about 1,700 years later, megafaunal extinction had begun and was quickly completed, so here it is attributed to a combination of climate change and human predation.[60]

The extinct megafauna of Australia are much more than the bull-sized marsupial tapirs, and included two even larger marsupials (notably the hippopotamus-sized *Diprotodon*), a number of kangaroos and wallabies – giants in comparison to their modern counterparts – and the ferocious tree-dwelling marsupial lion (*Thylacoleo carnifex*): 'the fellest and most destructive of predatory beasts', in the words of the nineteenth-century palaeontologist who gave it its scientific name.[61]

Australia also had many large birds (dromornithids) – taller and larger bodied than today's native emus and cassowaries – which have been extinct for millennia, including one (*Genyornis newtoni*) that may feature in Aboriginal artwork and stories (see colour plate section). As for the *bunyip* (see above), this possibility suggests that some of the extinct megafaunal species co-existed with people in Australia for some time, implying that the overkill model may not apply here. This is something of a departure from earlier views that envisaged most species of large-bodied mammals in Australia becoming extinct more than 40,000 years ago as a result of their rapid overkill by recently arrived humans.

The first data-informed suggestion that humans and megafauna had co-existed in Australia for at least 15,000 years came from measurements of rare-earth elements (REE) in megafaunal bones. The context of this research was that such bones had long been recognised in comparatively young sedimentary sequences, such as river terraces, but had been suggested as not being in situ. In other words, it was argued that these bones must have been moved (typically by flood waters) from the places where the animals died much earlier, to become incorporated within these much more recent landforms. If this scenario is correct, then the pattern of REEs in the buried bones would be expected to be quite different from that in the water filling the pores in the sediments enclosing the bones. Analysis of a whole series of megafaunal bones from Cuddie Springs in south-east Australia showed otherwise, leading to the conclusion that people (at least here) did not rapidly deplete Australian megafauna following their earliest encounters with them.[62]

This then raises the likelihood that it was climate change rather than humans – or a combination of factors – that drove megafaunal extinction in Australia. But the mechanisms are 'fiercely contested' and most recent scientific commentary still favours humans as the principal cause of this. A case in point is the timing of megafaunal extinction on Tasmania. While most megafauna on mainland Australia became extinct some 46,000 years ago, implicating expansionist humans, megafauna on Tasmania were once thought to have disappeared before humans reached this now-offshore island, exonerating human predation and implicating climate change. Yet improved dating of megafaunal remains from Tasmania tells a different story. It has shown that humans overlapped for just a couple of millennia with megafauna here, many of which they pursued across the land bridge then linking Tasmania with the Australian mainland, before exterminating them.[63]

It was several thousand years ago in southern Australia. Nothing stirred in the midday heat, the gum trees stood in desultory clusters. Within their foliage perched half a dozen Aboriginal hunters, unmoving, their barbed spears poised at arm's length ready to dispatch with maximum speed in an instant. Their scent made the giant flightless birds they hunted inquisitive yet unsuspecting. Slowly a group moved closer to the trees, within spearing distance. Standing over 2m (6½ft) tall, their 'heads … as high as the hills', these birds – *mihirung paringmal* to the Tjapwurung hunters – were feared for their vicious kicks, which could easily 'kill a man', yet prized for the huge amounts of meat even one would provide for the community. Closing on the trees in which the hunters waited, the birds peered upwards and, at the agreed signal, the hunters launched their spears in unison. The birds ran, faster than could any person, but those most severely wounded eventually collapsed and were dispatched by the pursuing hunters.[64]

It has been proposed that the *mihirung* was *Genyornis newtoni*, a flightless bird (averaging 275kg/606lb in weight) that was one of a number of huge-bodied dromornithids that once occupied most of Australia. Excavated *Genyornis* bones have been identified by Aboriginal informants as being from what they knew as *mihirung* and there are also rock-art images that appear to show this bird (see colour plate section).[65] If correct, this suggests that far from being an early casualty of human contact like many other species of Australian megafauna, *Genyornis* did in fact survive until comparatively recently. Debate rages. Most scientists do not believe there is any evidence for *Genyornis* having lived within the past 41,000 years, while others take a contrasting view.[66] The answer may lie in an egg.

Found typically in eroding sand dunes along Australian coasts, a distinct type of fossil eggshell has long captured the attention of palaeontologists. Reconstructed eggs – none have been found unbroken – typically have a long axis of around 16cm (6¼in), far larger than emu eggs, and are smooth as opposed to having the crenulated exterior of these. For some 30 years, it was generally believed that these eggs were those of *Genyornis newtoni*, and radiocarbon ages from them in every part of Australia have been used to argue a recent date for *Genyornis* extinction, perhaps just a few millennia ago. But then some nagging suspicions began to surface. For instance, the sizes of the reconstructed eggs are improbably small given the likely size of *Genyornis*. More recently, chemical analyses of the eggshells led researchers to conclude that they are 'unlikely to have been laid by a dromornithid, whereas several characters support a megapode origin'.[67]

There are several species of megapode living today in Australia. They either bury their eggs in natural deposits (like sand dunes), or build mounds where these are lacking. Giant megapodes once existed, too, and because of their characteristically outsize eggs would probably have become rapidly extinct upon making contact with egg-eating humans. So it seems that these eggs belonged to a giant megapode and not to *Genyornis*. The implications of this deduction are huge, for much of what has been previously inferred about the diet and time of extinction of *Genyornis newtoni* (mistakenly obtained from megapode eggshell) is now incorrect. This leaves open the question of whether *Genyornis* survived longer, even long enough to have humans recall how best it was hunted and to paint how it appeared.[68]

The most likely scenario for the extinction of Australian megafauna is that many species were in fact driven rapidly to extinction following early contacts with people, but that others, perhaps because of where they lived or how effectively they were able to dodge (and reproduce away from) human predators, survived much longer than the blitzkrieg 41,000 years ago. It is possible that some species of megafauna became extinct only within the past few thousand years, and that Aboriginal stories of the *bunyip* and kindred creatures may have their roots in observations of now-extinct megafauna. Of course this raises the possibility that the same may apply in other parts of the world.

The next chapter discusses ways of measuring other types of story (folk tales, for example), and where the new frontiers of knowledge of oral history may lie. It concludes by asking whether we have underestimated ourselves by denying that oral recollections can reflect human history across a far greater time span than we might intuitively suppose.

CHAPTER SEVEN

Have We Underestimated Ourselves?

By understanding that there are stories which have survived through largely oral means in our collective human psyche for several thousand years, not merely in isolated or anomalous situations, we are inevitably drawn to wonder whether in fact we have underestimated ourselves. As individuals, do we perhaps know stories that we have instinctively regarded as being of recent origin, but which are in fact far more ancient echoes of observations made – or stories invented – by our distant ancestors? Or as members of the human society, or cultural subgroups within it, do we have access to information – perhaps written, perhaps only oral recollection – that is of a considerable antiquity, significantly greater than we hitherto suspected? Today, the answers to both questions appear likely to be 'yes'.

We privilege the written word over the oral because, for most of us reading this book, this is what we have been uncritically trained to do. The culmination of our formal education was to learn essential knowledge through reading, not from our grandparents' rambling reminiscences of passing worlds. But consider that for most of the time – at least 90 per cent of the time in most of the world's cultures – humans have received knowledge only through the spoken words of others, usually our ancestors. Do we unquestioningly assume that all that knowledge was written down when our ancestors became literate? Even for a tiny proportion of that knowledge, this assumption seems improbable because, as numerous examples in this book demonstrate, some of the discoveries that we have made through scientific investigations and that we therefore regard as 'new' knowledge are in fact 'old' knowledge.

Our ancestors witnessed the rising of ocean levels after the end of the last ice age, and the often dramatic changes it wrought in coastal landscapes worldwide. In some places – Australia, north-west Europe and India – people successfully passed on their observations of coastal drowning for hundreds of generations into the age of literacy and hence down to us today. What science has discovered in the last 100 years or so about postglacial sea-level rise confirms the eyewitness accounts of our ancestors, not the other way round.

It is the same with stories of volcanic eruptions, meteorite impacts, and islands in the ocean basins that disappeared. Perhaps it is even the same in the case of now-extinct animals with which our ancient ancestors once rubbed shoulders. Yet because many such stories were first reported to a literate world, typically within the past 200 years, from the mouths of non-literate peoples, we have long regarded these stories as anthropological curiosities, expressions of oral cultures that are inherently inferior to the expressions of literate, science-informed cultures. That also seems to have been mistaken. With hindsight there is no doubt that science, especially in its younger days, could have benefited from treating oral traditions and knowledge more seriously. Some would no doubt argue that we might still benefit significantly from taking many non-literate sources of knowledge more seriously.

For example, Aboriginal Australians understood clearly that drought every few years led to massive bush fires in Australia, but that these might be contained in extent and ferocity by proactively reducing the amount of dry fuel load; current environmental management in Australia and elsewhere could indeed benefit from closer study of Indigenous practices. And in those societies where high levels of non-literacy exist today, there are numerous examples of how oral knowledge about particular phenomena – from the management of coastal risk in Kerala (India) to the recognition of tsunami and eruption precursors on islands like Simeulue (Indonesia) and Savo (Solomon Islands) – is favoured by local people over the (often inappropriate) solutions, underpinned by Western science, which many national or regional managers instinctively favour.[1]

Despite such positive examples of the usefulness of oral knowledge – and there are numerous others – this information source has also been long undervalued because its ancientness has rarely been able to be demonstrated. Unless we can demonstrate that a particular oral tradition is ancient, perhaps several thousand years old, rather than something a group of people invented a few weeks back, the argument often goes, why should we treat it seriously? With oral traditions, antiquity is a coarse proxy for veracity. And for such 'traditional' knowledge to be able to compete with scientific knowledge, it has to be verifiable – as scientific deductions generally are.

Science allows us to give credibility to many oral traditions. The main focus of this book – Australian Aboriginal stories of coastal drowning – involves the use of sea-level science to demonstrate that such stories in Australia must be at least seven millennia old. Radiometric dating allows us to know that Klamath stories of Mt Mazama date from around 7,600 years back. And so on.

Ways of measuring the antiquity of other types of story have also been developed recently and show great promise for the future.

A little girl in a hooded red jacket walks, somewhat apprehensively, through the woods to her grandmother's house, clutching a basket of bread and apples. But instead of finding her grandmother home, she encounters a wolf in her bed, disguised, and – depending on the version of the story – she is either eaten or rescued and the wolf is killed. Known to English readers as *Little Red Riding Hood*, the earliest written version of the story was *Le Petit Chaperon Rouge* in 1697 by Charles Perrault, widely credited as the first writer of fairy tales. Yet as Perrault happily acknowledged, the sources for his fairy tales – which also included the first written accounts of *Cinderella*, *Puss in Boots* and *Sleeping Beauty* – lay in popular oral stories that had been told for generations in Europe. But for exactly how many generations? Can we know?

We can indeed find an answer to this question by applying to stories the same kind of tools that biologists have long used for tracing the evolution of species. This process involves identifying diagnostic characteristics of a range of species, deciding which characteristics and what proportion of these are shared between species at several points in time, then drawing up a phylogenetic tree to illustrate the likely process of species evolution. An analogous process is used to analyse the evolution of stories like *Little Red Riding Hood* and thereby pinpoint the time (and sometimes place) of their earliest expression. In short, the process begins with identifying key motifs within different versions of a particular story and those in others similar to it; such motifs might be the existence of a fairy godmother, the use of a magic ring or even the activities of a giant kangaroo.[2] Any combination of motifs that tells a story is a 'tale type', and it is tale types that provide the raw material for phylogenetic analyses of oral–written stories.

The *Little Red Riding Hood* story and its close relative, *The Wolf and the Kids*, have been found in 58 versions from numerous cultural traditions in Europe, but also in Africa and East Asia. On the assumption that the geographical concentration of tale types sharing the greatest number of defining motifs is likely to be the source of these tales, the indication is that this particular story originated as an oral tradition in the eastern Mediterranean at least 2,000 years ago, and was first written down in Greece as a fable in about AD 400. From here, the story spread along the Silk Road to China, across the Mediterranean to Africa, both orally and in writing, evolving different characteristics as it was adopted and reframed by other cultures.[3]

Phylogenetic analysis of such stories, like many of those quoted earlier in this book, holds great potential for understanding the ways in which particular tales evolved, and when and where they are likely to have originated. But whereas such analysis focuses on trying to understand the ways in which particular tales have changed across the centuries and millennia that people have been telling (or reading) them, most of the Australian Aboriginal stories recounted here appear to have changed so little that their commonalities are readily detectable, whether they refer to coastal drowning or to volcanic eruptions and kindred phenomena.

In the cases of both the *Little Red Riding Hood* fairy tale and Aboriginal stories about the time, perhaps 10 millennia ago, when the Great Barrier Reef was the coastline, recent research has demonstrated with a high degree of plausibility that humans can pass on memories for several thousand years without the assistance of literacy. This, I feel, should make scientists of many different hues sit up and consider whether we have underestimated ourselves – whether, caught up in the whirlwind of literacy and all its attendant possibilities, we have discarded something valuable – its antecedents.[4]

So where is the edge of memory, the point in time beyond which humankind has no direct record?

It lies probably more than 10 millennia ago but perhaps nowhere near 20. It is not that a case cannot be made for the latter, but simply that applying the law of parsimony – which tells us to choose the simplest scientific explanation of the facts – makes it improbable.[5] In this book, examples of oral memories of events that apparently occurred more than 10,000 years ago have been related, but the precise antiquity of many is difficult to assess, either because the event (like an eruption or a landslide) has itself proved difficult to date, or because the errors inherent in many dating techniques unavoidably reduce precision.

Consider the first of these points. With many 'drowning stories', for instance, the depth below the present sea level to which a particular story refers is often difficult to fix. For example, do the ubiquitous Aboriginal stories of the time when the Great Barrier Reef was the coastline refer to a sea level 10, 45 or even 65 metres below the present sea level? It makes a huge difference to the minimum age of these stories. And if a lava flow is too weathered to date directly, how close are the 'minimum' or 'maximum' ages obtained from proxy eruption indicators to its formation? It also makes a difference.

Then you have the second point, the fact that most dating techniques are imprecise. Radiometric dating techniques produce results that are expressed as age ranges, a function of the techniques used. Other techniques, such as linking drowning stories to envelopes showing how sea levels changed in the past, often amplify those kinds of error.

But these are caveats. They should not deter us from pursuing ways of measuring the antiquity of oral traditions, of calibrating the longevity of memory.

Stories of coastal drowning from five places in Australia appear (see Table 4.1) to date from more than 9,000 years ago, and perhaps over 13,000 years ago. Of these five, the most compelling are likely to be those from Spencer Gulf (9,330–12,460 years ago) and Kangaroo Island (10,080–10,950 years ago). Elsewhere in the world, drowning stories of similar antiquity may be preserved for Baie de Douarnenez in France (perhaps 8,750–10,650 years ago) and Cardigan Bay in Wales (perhaps 9,000–10,250 years ago). There are many other drowning stories dating from more than 6,000 years ago.

Then there are other types of story, including the Australian Aboriginal memories of the Tyrendarra lava flow from Mt Eccles (more than 10,500 years ago) and the formation of the *maar* volcano at modern Lake Eacham (9,130 years ago), the explosive eruption and caldera formation at Mt Mazama in the western United States (7,600 years ago), and perhaps even the landslide-damming of the Citarum River at Tangkuban Perahu in Indonesia (16,000 years ago?).

Together, these data imply that human memories can remain alive for many millennia. Others, as yet undocumented, perhaps exist. We should consider the possibility.

As a final point in this book, consider that if the edge of our memories today lies 10 millennia or so in the past, what of people who lived earlier than this, in those times beyond the edge of memory?

Our species, modern humans, has been in existence for almost 200,000 years. For much of this time, oral communication has been the main form of knowledge transmission between individuals. But what happened to knowledge that was more than 10,000 years old, which slipped into the abyss over the edge of memory? It became forgotten – just as today much of our ancestors' knowledge has been forgotten. Was that forgotten knowledge important? Did things have to be rediscovered or reinvented because of that knowledge loss? Absolutely.

In our species' history, there are many examples of innovations that did not endure, ideas that probably made life easier for a time but did not last. One of the most obvious refers to the ability of humans to cross water gaps too large to swim. Consider the first people to arrive in Australia, who had to negotiate ocean gaps as much as 70km (43 miles) wide. Obviously they used boats or rafts that were able to make the crossing, a sufficiency involving not only boat-building technology but also maritime and navigational skills. Yet 65,000 years or so after they reached Australia, their descendants had no such skills, and were hardly able to sail far from shore in most cases.[6] What happened? Was it simply a case of no longer needing an ability to travel long distances across the ocean, or was it that people just lost the knowledge of how to do so?

It seems that the history of our species has been marked by an alternation between knowledge acquisition and knowledge loss through gradual memory attenuation. It is probable that in societies that had to cope with especially harsh environments, like those of the Aboriginal Australians, optimally effective techniques for intergenerational knowledge transmission were evolved. Today we look at those societies in awe, as examples of how it might once have been for all of us.

Notes

References in these notes are given only to the books listed under Further Reading. Where reference to other sources, typically scholarly articles, cannot be avoided, only their lead author, the year of publication and the journal name (in italics) are given.

Chapter 1: Recalling the Past

1   Hillman described his discovery in the *Portland Oregonian* newspaper on 3 June 1903. More details at www.craterlakeinstitute.com.

2   p. 36 in Deur (2002, *Oregon Historical Quarterly*).

3   Quoted on p. 39 of Deur (*op cit*).

4   You may wonder why the example of the tsunami of 27 March 1964 that impacted many parts of the Cascade coast is not used. It is because even though the earthquake that caused it occurred above a zone of plate convergence, it was generated off the south coast of Alaska, not locally.

5   A comprehensive account of the 1700 tsunami, nicknamed the 'orphan tsunami' because the earthquake that caused it has not been identified, is provided in the book by Brian Atwater and colleagues (2005), which painstakingly pieces together its effects along the continental margins of the North Pacific.

6   The parallels between this story and that of Persephone in Greek mythology are striking yet must be coincidental. Like Loha, the beautiful Persephone lived on the Earth's surface but was abducted by Hades, the God of the Underworld, who was in love with her (and up against some stiff competition). Hades kept Persephone in his underworld kingdom until ordered by Zeus to return her to the above-ground world. But Hades had ensured that Persephone had tasted the food of the underworld (pomegranate seeds), something that compelled her to return there to his cold embrace for several months each year.

7   From pp. 53–55 of Clark (1953). The most convincing
    interpretation of this story is in the magnificent book *When
    They Severed Earth from Sky: How the Human Mind Shapes
    Myth* (Barber and Barber 2004).

8   The most convincing dates come from the tephra (ash
    deposits) blown out of Mt Mazama across a wide area during
    its terminal eruption. Radiocarbon dating of associated
    materials, such as wood fragments and rat faeces, within the
    tephra deposits shows that they were laid down between
    7,682 and 7,584 calendar years BP (Before Present, where
    'present' is AD 1950) (Egan, 2015, *The Holocene*).

9   This quote and the other below it come from the translation
    of the key oral history by Beaglehole and Beaglehole (1938:
    386). Around 85km (53 miles) from the nearest inhabited
    land – the equally remote Nassau Island – Pukapuka is 175ha
    (432 acres) in area and rises mostly less than 2m (6½ft) above
    sea level. Some 600 people live there today.

10 The original text is on p. 116 in the book by Ricci (1969).
    The translation given here is on pp. 316–317 in the paper by
    Santilli and others (2003, *Antiquity*). Other quotes in this
    section come from the same source. Note that the calibrated
    radiocarbon age for the impact (AD 412) is unlikely to be its
    exact age.

11 It is possible that the Dionysian temple referred to was
    converted to the Church of Santa Maria della Consolazione
    at Secinaro.

12 The oldest known fossils of *Homo sapiens* are found in
    Ethiopia and are consistent with genetic evidence suggesting
    the emergence of our species about 200,000 years ago in the
    region (Gibbons, 2003, *Science*). See also Chapter 4.

13 Our hominid ancestor, *Homo erectus*, moved out of Africa,
    where it had evolved, to reach both China and Java more
    than 1,600,000 years ago. While *H. erectus* did not have the
    cognitive abilities of our species, its presence on Flores Island
    (Indonesia) at least 800,000 years ago shows that it was able
    to cross ocean distances of at least 19km (12 miles)
    (Morwood, 1998, *Nature*).

14 Of the clam species living today along the shores of the Red
    Sea, *Tridacna costata* comprises fewer than 1 per cent. Yet this
    species comprises about 80 per cent of the fossil shells found
    in the ancient shoreline of the region, demonstrating that it
    was once far more common. The association between these
    fossil shells of *T. costata* and human artefacts shows that its
    predation by people began at least 125,000 years ago.

15 Owing to the comparative abundance of food resources along
    the coasts of South Asia, it is generally thought that the
    dispersal of *Homo sapiens* east out of Africa followed a coastal
    route, most likely between about 70,000 and 130,000 years
    ago.

16 Perhaps they inferred that land existed because they saw
    smoke rising from distant wildfires. Perhaps they observed
    birds flying south and not returning, from which they
    deduced that a sizeable land mass with abundant resources
    existed in that direction.

17 Beyond the scope of this book is the question of what came
    before language. It was probably gestural communication –
    based on signs – that may have evolved into speech through
    echo phonology, 'a repertoire of mouth actions which are
    characterized by "echoing" on the mouth certain of the
    articulatory actions of the hands' (Woll, 2014, *Frontiers in
    Psychology*, p. 1).

18 Ever insightful, it was Jonathan Kingdon who suggested in
    his book *Self-made Man* (1996) that, once humans reached
    continental shores, they were no longer able to hunt and
    gather their food at the cooler (low sun) times of the day they
    preferred. Instead they became dependent on the tides, even
    sometimes having to walk out across reefs collecting shellfish
    at the hottest times of the day. Over time, people with darker
    skins became more successful at this, and less likely to suffer
    sunstroke or contract skin cancer, thereby resulting in a
    gradual darkening of human skin colour in parts of the world
    like coastal South-east Asia 70,000 years ago. Kingdon names
    the first people of this area with black (rather than brown)
    skins, the Banda, and speculates that their seafaring skills may
    have started with the building of rafts on which seafoods

could be piled, then progressed to platforms on which people could sit, and eventually to mobile watercraft.

19 The Wallace Line separates Bali from Lombok, Borneo from Sulawesi. On the Asian side (commonly called Sunda) are found land animals like orang-utans and rhinoceroses that are not found on the Australasian (commonly called Sahul) side, where an abundance of unique fauna (83 per cent of it endemic) – including kangaroos and koalas – exists.

20 The most efficient way that people could have crossed from the now-submerged extension of the Asian continental shelf to that of the Australian shelf would have been from south-east Borneo across Sulawesi Island to New Guinea, a route requiring eight ocean crossings, the largest being 70km (43 miles); from New Guinea at that time, it was a dry-foot walk to modern Australia. The earliest known date for a human presence in Australia and New Guinea comes from the Malakunanja rock shelter on the side of the valley of the East Alligator River in the Northern Territory. This rock shelter was occupied 52,000–61,000 years ago, but, being some 220km (137 miles) from the coast at this time, is likely to have been reached at least a millennium or two after the first human arrivals. More details in Chapter 2.

21 Many Aboriginal Australians reject 'Western' scientific ideas about their ancestors 'coming' to Australia from anywhere. Like many indigenous peoples, a belief that they have always 'been here' [where they are now] is commonplace. An insightful biographical extract, quoted by van den Berg (2002: 4), reports that:

> As a child, around campfire talks with my parents … I would ask, 'Where did we come from?'
> Their reply was, 'We've always lived here.' I accepted that explanation.
> As I grew older and matured, I would ask the same question and get the same answer. 'We've always lived here.'
> Later in my learned wisdom of European ways, I would reply, 'But white fellas say we come from overseas, from Asia.'
> 'Well,' they said, 'Those *wujbullas* are talking out of their *nooroos* [their backsides]. They don't know anything.

Our Dreamtime stories tell us our Rainbow Serpent made us and our land, our Mother. We belong here, to this country … and don't you forget it.'

It is worth reflecting that there is no direct evidence that anyone 65,000 years or so ago crossed 70km (43 miles) of ocean between South-east Asia and Australasia – although the weight of inference suggests that they did.

22 I am aware this paints a fairly idealistic picture, fine for the purpose of generalisation but unrealistic when applied to the evolution of particular societies. When considering how societies diverged (and came to speak different languages) in the past, a process that can be tracked by genetics, it is often thought that language facilitated knowledge dissemination and that inequalities which appeared between different groups were a result of their respective histories. But it seems more likely that language was not used to freely communicate knowledge but rather to selectively *withhold it*, thereby creating a situation in which one group was better able to survive than another (Iain Davidson and Bill Noble, 1992, *Archaeology in Oceania*).

23 Urban centres are defined here as settlements large enough and complex enough to support people having roles other than those of primary food producers. Dates of around 7,000 years ago for the establishment of complex 'villages' in the Yangtze lowlands may qualify these as urban, although 1,000 years later in the same area, supported by rice agriculture, walled and moated cities were built.

24 Mesopotamian cities of note include Tell Brak (Syria), first established about 7,000 years or so ago. Recent research suggests that urbanisation in this so-called Fertile Crescent was a 'phased and pulsating phenomenon' rather than a continuous process (Lawrence, 2015, *Antiquity*). The agro-pastoral strategies were centred on the floodplain cultivation of wheat, barley and lentils, with olives and grapes in wetter areas complemented by domesticated food animals like pigs, then increasingly by sheep and goats as time went on. The latter also facilitated wool-based textile production in the region.

25 The evolution of imagination (and the associated vocabulary) is a subject of some debate. Humans may be the only extant species that uses imagination to help process things we perceive (Bronowski 1974). There is debate about whether the human imagination is innate (perhaps stimulated by dreams) and indeed defines us as a species, or whether it derives solely from the reception (typically through observation) and processing of external stimuli. At the moment, I favour the latter and draw an explicit link between our ancestors' observations of memorable natural phenomena (like volcanic eruptions, meteorite showers and giant waves) and mythmaking that underpins much of today's creative practice, including art and literature.

26 The subject of a notable compilation by Dorothy Vitaliano (1973).

27 My own research on the island of Kadavu (Fiji) highlights just such a situation (see also Chapter 6). At one time, the forest-cloaked volcanic mountain named Nabukelevu at the western extremity of Kadavu was assumed to be extinct, last erupting perhaps 50,000 years ago. But then there are myths from the nearby island of Ono that could be interpreted as observations of eruptions of Nabukelevu. Since people have been in Fiji for only about 3,000 years, these myths imply that the volcano is unlikely to be truly extinct. More recently a road cut along the base of Nabukelevu revealed pottery fragments covered by volcanic scoria – a sure sign that the most recent eruptions postdated human arrival in the area – and a validation of the myth.

28 These include the Big Island (Hawai'i) in the Hawaiian Islands, which is the highest single mountain on Earth, reaching more than 10km (6 miles) above the surrounding ocean floor. Through its Kilauea parasite, Hawai'i has been erupting continuously since 1983.

29 I analysed the Maui legends of the Pacific Ocean and explained how their foci appear to coincide with recent shallow-water volcanism (Nunn, 2003, *Annals of the Association of American Geographers*).

30 In his landmark 1982 book *Orality and Literacy*, Father Walter
    Ong wrote that 'oral cultures indeed produce powerful and
    beautiful verbal performances of high artistic and human
    worth, which are no longer even possible once writing has
    taken possession of the psyche. Nevertheless, without
    writing, human consciousness cannot achieve its fuller
    potentials, cannot produce other beautiful and powerful
    creations' (p. 14).

31 The argument is that at one time most human females died
    shortly after they reached childbearing age, but when an
    increasing number began to survive into older age and
    became grandmothers, they (and to a lesser their male
    counterparts) became key in imparting traditional knowledge
    to their grandchildren (Caspari, 2004, *Proceedings of the
    National Academy of Sciences of the United States of America*).

32 This is the mnemonic effect, elegantly described by Lynne
    Kelly (2016).

33 Traces of San ancestors – including spearheads, notched
    bones, warthog tusks and ostrich eggshell beads – have been
    found in Border Cave, South Africa, and dated unequivocally
    to somewhere between 45,000 and 60,000 years ago.

34 p. 243 in Guenther (2006, *Journal of Folklore Research*).

35 p. 113 in the chapter by Hough in a collection on *Myth and
    Geology* (Piccardi and Masse 2007).

36 p. S79 in Hunn and others (2003, *Current Anthropology*).

37 p. S87 in Hunn and others (*op cit*). The Tlingit had several
    other notable conservation practices designed to maintain
    supplies of particular food resources, including an avoidance
    of the use of paralytic fish poisons (used by many other
    cultural groups) that have unintended effects on many other
    nearshore marine food sources, particularly shellfish.

38 This is based on research at the Ozette archaeological site on
    the Washington coast, where it was found that fur-seal
    populations had not been discernibly affected by human
    predation in the period AD 1100–1800. It should be pointed
    out that not all such interactions between indigenous peoples
    and wild foods were so apparently sustainable. There are
    innumerable examples of instances where hunger evidently

took precedence over any sense of sustaining a future supply of a particular food resource.

39 In his poem 'The World-Soul', Emerson compared the present where 'the politics are base' and 'the letters do not cheer' with a place 'far in the depths of history' where we find that 'voice that speaketh clear'.

40 Two pairs of woollen trousers (long pants) made and worn about 3,000 years ago have been excavated from an archaeological site at the Turfan Oasis in western China (Beck, 2014, *Quaternary International*). It is likely that they were invented to allow horseback riders to remain comfortably for longer in the saddle, something that allowed more efficient and widespread communication at the time.

Chapter 2: Words that Matter in a Harsh Land

1 We do know that water availability was the single most important control on early (Aboriginal) exploration of Australia. Eighty-four per cent of archaeological sites more than 30,000 years old in Australia are within 20km (12 miles) of permanent water. And there were several adequately watered routes into the arid interior of the continent available to potential Aboriginal settlers (Bird, O'Grady and Ulm, 2016, *Proceedings of the National Academy of Sciences*).

2 The semi-arid lands in much of the Murray-Darling Basin contain huge quantities of salt that in places wash into the rivers and the groundwater. Not derived from bedrock (as you might expect), this salt is of largely aeolian origin, carried to the area by winds from the ocean and deposited in clay mantles across the ground surface. Today the extraction of salt from groundwater allows water better suited for agriculture to reach productive areas downstream, and also produces crystalline salt in commercial quantities.

3 The quote is from Sturt (1834: 108). I cannot resist adding Sturt's story of his encounter with kangaroo flies: 'We remained stationary the day after we left the range, with a view to enjoying a little rest; it would, however, have been infinitely better if we had moved forward. Our camp was infested by the kangaroo fly, which settled upon us in

thousands. They appeared to rise from the ground, and as fast
as they were swept off were succeeded by fresh numbers. It
was utterly impossible to avoid their persecution, penetrating
as they did into the very tents. The men were obliged to put
handkerchiefs over their faces, and stockings upon their
hands; but they bit through every thing. It was to no purpose
that I myself shifted from place to place; they still followed, or
were equally numerous everywhere. To add to our
discomfort, the [pack] animals were driven almost to
madness, and galloped to and fro in so furious a manner that I
was apprehensive some of them would have been lost. I never
experienced such a day of torment; and only when the sun
set, did these little creatures cease from their attacks' (p. 71).

4 Australia is currently moving north-north-east at a rate of
7cm (2¾in) each year, so fast (in geological terms) that the
geocentric datum of Australia, defined by global latitude and
longitude coordinates in 1994, was found in 2016 to be out
by more than 1.5m (5ft) – see report accessed in September
2016 at www.abc.net.au/news/2016-07-28/
aust-latitude-longitude-coordinates-out-by-1-5m-
scientists/7666858.

5 A billion is 1,000,000,000. By measuring its U-Pb (uranium-
lead) proportions, a zircon grain from the Jack Hills was
dated to 4.4 Gyr (4,400,000,000 years) ago, 'shortly after
formation of the Earth' (Valley, 2014, *Nature Geoscience*,
p. 222).

6 It has been proposed that these ancient stromatolites
developed in very similar conditions to those found on the
surface of the planet Mars, and that they may in fact exist
there.

7 Probably the last word on Australian geology for many
decades to come is the beautifully illustrated and explained
volume *Shaping a Nation: A Geology of Australia*, published in
2012 to coincide with the 34th International Geological
Congress in Brisbane (Blewett 2012). It is the heaviest single
book in my library.

8 Australia is currently home to more camels than any other
country on Earth. Introduced in the 1800s to work in the

country's deserts, initially for their exploration and later for supply trains to aid infrastructure development, feral camel herds have grown vastly and are now subject to periodic culling.

9 In the words of one nineteenth-century traveller, the stony deserts of Australia's interior are 'so desolate that it is horrifying even to describe [them] ... truly the wanderer in its wilds may snatch a fearful joy at having once beheld the scenes, that human eyes ought never again to see' (from Book 5 of Giles 1889: 317–318).

10 Quotes from an unpublished letter by Charles Sturt.

11 Properly Malakunanja II Rockshelter, also called Madjedbebe.

12 These dates were obtained using optically stimulated luminescence (OSL), a measure of how long an artefact has been buried (and unaffected by ionising radiation), and are actually $53,400 \pm 5,400$ to $60,300 \pm 6,700$ years ago (Roberts, 1994, *Quaternary Science Reviews*).

13 The Malakunanja artefact ages are actually $45,000 \pm 9,000$ to $61,000 \pm 13,000$ years ago (Roberts, 1990, *Nature*). These ages were also obtained using OSL (see previous note), and any doubt that they might be too old, perhaps because the grains of material being analysed had received insufficient exposure to sunlight (and ionising radiation) before being buried, was dispelled by subsequent analysis.

14 There are numerous studies that trace the evolution of stone-tool manufacture in Australia through time (e.g. Mulvaney and Kamminga 1999).

15 The age quoted for the Layer 30 hearth at Devil's Lair was reported by Chris Turney and others (2001, *Quaternary Research*).

16 The mean age of this individual is $62,000 \pm 6,000$ years, so they could have been living as much as 68,000 years ago (Thorne, 1999, *Journal of Human Evolution*).

17 Research on the mitochondrial DNA (mtDNA) of ancient Australians – including Mungo Man – was once thought to indicate that there was a race of people in Australia before Aboriginal people arrived, a suggestion that more recent work

has conclusively dismissed (Heupink, 2016, *Proceedings of the National Academy of Sciences*).

18   Actually radiocarbon dates on charcoal can mislead in some instances. This may be because the particular tree that was burnt lived a long time – perhaps several hundred years – so that the age determined for it may be out by this much. And then it is always a possibility that the people who burnt wood, particularly for cooking fires or in ceramic kilns, used wood that was long dead – like driftwood on a beach – so that the calculated age might also be significantly different from the time of burning. These caveats apply most commonly to younger charcoals, where a few hundred years might make huge differences to the interpretation of human prehistory, but much less when we are dealing with older, less precise radiocarbon ages.

19   Such 'fire stick farming' was a hallmark of Australian Aboriginal subsistence strategies. Not only did it reduce the build-up of (vegetation) fuel loads and drive away venomous snakes, but it subsequently stimulated the growth of new plants, including edible bracken, which attracted animals like kangaroos that could then be more easily hunted.

20   The original interpretation of early human agency was based on analyses of charcoal and grass pollen in Ocean Drilling Project (ODP) core 820, but revised subsequently when it became clear that El Niño–Southern Oscillation (ENSO) effects could produce the same kinds of change.

21   The 120,000-year date comes from the original study by Gurdip Singh and others in a chapter in the collection by Gill and others (1981).

22   Re-dating of Zone F suggests that its correct age is about 60,000 years old, in line with the earliest dates for initial human occupation of other parts of Australia. Still, it is possible that Zone F was produced by natural processes, uninfluenced by human activity.

23   Waisted stone axes are considered to represent a huge technological jump in stone-tool design that opened up new possibilities, particularly for forest dwellers, as most people in Papua New Guinea were 40,000 years ago.

24   Many archaeologists and others would favour a younger age,
      pointing out that the 60,000-year ages from Malakunanja
      and Nauwalabila are at the upper ends of the OSL ages for
      these occupations, and that there is evidence – discussed in
      Chapter 6 – which implicates humans in an Australia-wide
      mass extinction of megafauna within a few millennia
      beginning some 45,000 years ago. The most recent research
      at Malakunanja appears to confirm an age for the site's
      occupation in excess of 60,000 years ago (Clarkson, 2017,
      Nature).

25   The key study is that by Heupink (*op cit*), who used improved
      DNA sequencing methods to re-evaluate the quite contrary
      yet influential conclusions of an earlier study. The latter
      analysed mitochondrial DNA (mtDNA) – that which is
      passed through the maternal line – from 10 skeletons found
      in western New South Wales that dated from well before the
      European arrival in Australia in 1770. By comparing the
      mtDNA sequences in these individuals to those from modern
      Aboriginal Australians, it was found that most had much in
      common but that – critically – two of the skeletons contained
      mtDNA that was *not* diagnostic of Aboriginal people. It was
      therefore concluded that there had not been just one period
      of human arrival into Australia before 1770, but multiple
      waves that explained the genetic diversity. The more recent
      study re-examined much of the original skeletal material as
      well as some new material. The conclusions were staggering.
      Some of the original bones analysed were found not to
      contain any verifiable human DNA, begging the question as
      to how this might have allegedly been sampled 15 years
      earlier. But more importantly, it was clear that the
      indiscriminate handling of the skeletal remains by
      (non-Aboriginal) scientists had introduced some of their
      mtDNA to the mix, resulting in the non-Aboriginal mtDNA
      identified in the earlier study. The presence of an authentic
      Australian Aboriginal haplotype in at least one of the
      skeletons analysed proved enough to confirm that Aboriginal
      Australians have been isolated for most of the time – perhaps
      65,000 years – that they have occupied the continent.

26 The figure of 14,000 years ago for the sundering of the last
    land connection between mainland Australia and Tasmania
    comes from research by Lambeck and Chappell (2001, *Science*).
27 The research was based on mitochondrial DNA and suggests
    that dingoes were introduced to Australia from Indonesia
    (Oskarsson, 2012, *Proceedings of the Royal Society*).
28 A possible introduction from India to Australia during this
    event was the microlith, a small stone tool used in
    arrowheads and spear-tips. Until the genetic studies showing
    contact with people from India about 4,230 years ago, the
    abrupt development of microliths in Australian cultures
    around this time had puzzled archaeologists.
29 A study of the Macassan trepang-processing station at Malara,
    Anuru Bay, Arnhem Land (northern Australia) found
    numerous remains of earthenware pottery, introduced from
    about AD 1637, which all derived from the port of Makassar
    in southern Sulawesi. Macassans may also have introduced
    the tamarind tree to Australia. They certainly introduced
    numerous words that became part of Aboriginal languages
    along the continent's northern fringes, and are probably
    responsible for integrating aspects of Islam into Aboriginal
    spiritual beliefs and rituals in these parts of Australia.
30 One of the main purposes of Kerguelen's voyages in the
    Indian Ocean was to try and relocate 'Gonneville', a lush
    land apparently described from the region by Paulmier de
    Gonneville in 1504, but which subsequent research has
    determined to have been Brazil. Kerguelen's premature
    enthusiasm for Kerguelen Island may have arisen from his
    belief that it was actually Gonneville.
31 The rock art on sandstone cliffs in the Serra da Capivara
    National Park in Brazil has been dated to far earlier than the
    time at which it is generally assumed people first arrived in the
    Americas. Archaeological research suggests that people were in
    this area more than 20,000 years ago, which would mean that
    the orthodox view of the first Americans arriving from
    easternmost Asia across the Bering Strait to create the Clovis
    culture in the modern USA about 14,000 years ago, has to be
    wrong. This is not a popular view in some quarters, as you

might imagine, and has led one group of non-partisan archaeologists to wonder whether 'there is some sort of curse that affects the common sense both of archaeologists making the discoveries, and their colleagues, at the announcement of an age older than 14 000 cal BC' (Boeda, 2014, *Antiquity*, p. 928).

32   These types of wasp construct mud nests that petrify after they are abandoned. The example referred to was built by wasps of the species *Sceliphron laetum* and came from a rock shelter near the King Edward River Crossing in the Kimberley.

33   *Tjukurrpa* is the word of the Pitjantjatjara people of central Australia for the Dreaming and is widely used as a standard translation. Other words for the Dreaming are *Ungud* (Ngarinyin people of the Kimberley), *Wongar* (Yolngu people of Arnhem Land), *Altyjerre* (Arrernte people of central Australia), *Bulurru* (Djabugay people of north-east Queensland) and *Kulhal* (Yaraldi people of South Australia).

34   From p. 357 of David Rose's insightful 2013 article on the phylogenesis of the Dreamtime, published in *Linguistics and the Human Sciences*.

35   The range of books in the Dreaming Library were reported in much the same way in 1906 by the French ethnographer Arnold van Gennep, who found that they included '*des fragments d'un catéchisme, d'un manuel liturgique, d'un manuel d'histoire de la civilisation, d'un manuel de géographie, mais beaucoup moins d'un manuel de cosmographie*' (1906: cxiv), translated as 'fragments of a catechism, a liturgical manual, a history of civilisation, a geography textbook, and to a much lesser extent a manual of cosmography'.

36   'This kind of system, often labelled in Aboriginal English as the "owner-manager" relationship, requiring a story to be discussed explicitly across three generations of a patriline, constitutes a cross-generational mechanism which may be particularly successful at maximising precision in replication of a story across successive generations' (Nunn and Reid, 2016, *Australian Geographer*, p. 40). If you are unclear still, then imagine three generations A, B and C. Grandfather A teaches stories to his son B, who teaches them to his son C. B's sister gets her children Y (who learnt the stories from their paternal

grandfather X) to talk to A about B's and C's knowledge of the stories. Thus A, B and C all discuss the stories with Y, whose role is to 'manage' the stories within three generations of his mother's patriline. It works in the other direction as well. X, Y and Z all discuss their knowledge with B, who also has a role as a knowledge manager. Thus there is formalised cross-checking of stories across three generations of two patrilines. When a grandfather (A or X) dies, the role of knowledge manager also passes on; in the case of patriline A–B–C, which is managed by Y, the death of A will mean that patriline B–C–D is now managed by Z (Y's child). At each generational stage, an 'owning' patriline has its stories checked by a 'managing' maternal relative – the 'owner-manager' relationship. I am grateful to Dr Nick Reid for this illustration of how cross-checking of Aboriginal knowledge transmission across generations worked.

37 Australian Aboriginal performances that combine song, dance and performance are known as *corroboree*(s).

38 For Aboriginal stories, it is likely that 'song rhythms and tunes have a conventionalizing effect on the transmission of ideas in song form' (Berndt and Berndt 1996: 387), ensuring that detail is not changed or altered as stories are told anew to successive generations. Many of us today remember tunes we last heard a decade ago before we recall the associated lyrics.

39 The point is made neatly by Elizabeth Cameron, who wrote 'Australian Aboriginal symbols are visual forms of knowledge that express cultural intellect. Being classified by a Western interpretation of "art" devalues thousands of years of generational knowledge systems, where visual information has been respected, appreciated and valued' (2015, *Arts*, p. 68). She goes on to explain that 'Aboriginal creativity' was never considered by its creators to be primarily aesthetic, but rather in service to pragmatic ends.

40 The story of this map is told by anthropologist and linguist Donald Thomson and refers to a time when he was camped with the Pintupi at Labbi-Labbi on the edge of the Sandy Desert in 1957. 'And on the eve of our going, [Aboriginal informant] Tjappanongo produced spear-throwers, on the

backs of which were designs deeply incised, more or less geometric in form. Sometimes with a stick, or with a finger, he would point to each well or rock hole in turn and recite its name, waiting for me to repeat it after him. Each time, the group of old men listened intently and grunted with approval – "Eh!" – or repeated the name again and listened [to me] once more. This process continued with the name of each water hole until they were satisfied with my pronunciation, when they would pass on to the next. I realized that here was the most important discovery of the expedition – that what Tjappanoŋgo and the old men had shown me was really a map, highly conventionalized ... of the waters of the vast terrain over which the Bindibu [Pintupi] hunted' (p. 62 in Thomson, *Geographical Journal*).

41 Many European Australians in the nineteenth century found it difficult 'to appreciate that Aboriginal hunter–gatherers, whose material culture seemed to them so primitive, had a sophisticated artistic life, not to speak of one of the most complex systems of social organization in the world, and a religious life to which the older and more privileged members of society devoted a great deal of their time' (Ross, 1986, *Oral Tradition*, p. 232).

42 There was 'a comforting colonial conceit that the Aborigines made no use of their land' (Henry Reynolds in Foreword to Gammage 2011: xxiii).

43 The quote is from p. 81 of Tim Flannery's influential (1994) book *The Future Eaters*, and was the first popular expression of this fact – and of the decades of its increasingly obvious nature being ignored by settlers. Flannery also makes the observation that the nomadism practised by Aboriginal Australians was 'an adaptation to tracking the erratic availability of resources as they are dictated by ENSO' (p. 283), something manifested in the characterisations of Australia as a land of 'droughts and flooding rains', a place where long-term averages are consequently better represented by medians rather than by means (see colour plate section).

44 A good account of the two responses to drought stress by the
    Ngaatjatjarra (Ngatatjara) Aboriginal people of the Western
    Desert was given by Gould (1991, *Oceania*). 'Drought escape'
    involved the temporary abandonment of the worst drought-
    stricken areas and resettlement in distant ones where more
    water was available, often areas where kinfolk lived. 'Drought
    evasion' was often employed when the likely duration of a
    particular dry spell was considered shorter rather than longer;
    it involved the relocation of groups within their home area,
    usually congregating in places where there was a relatively
    dependable water supply. Gould argues that such drought-
    response strategies may elsewhere have been key to
    stimulating the domestication of plants.
45 See Cook (1893: 244).
46 Observation made by Sydney Parkinson (1773: 124) on 27
    April 1770.
47 This account is from pp. 136–137 of Haygarth's (1861)
    Recollections of Bush Life in Australia.
48 From pp. 412–413 of Thomas Mitchell's (1848) *Journal of an
    Expedition into the Interior of Tropical Australia*.
49 The assumption that Australia was in fact *Terra Nullius* in
    1770 when James Cook first saw it was overturned in 1992 by
    the High Court of Australia that ruled – by a six-to-one
    majority   that the Meriam people, who had brought the
    case, were 'entitled as against the whole world to possession,
    occupation, use and enjoyment of (most of) the lands of the
    Murray Islands'. This ruling opened the way for the
    introduction of native title into Australian law and has led to
    numerous instances of title being restored to groups of
    Aboriginal landowners.
50 Quote from Flannery (1994: 282). While nomadism was
    characteristic of most Aboriginal groups until post-
    colonisation, there is clear evidence that some groups that
    found themselves stranded by postglacial sea-level rise on
    offshore islands responded by both increasing their
    consumption of coastal-ocean foods and becoming more
    sedentary. This transition is demonstrated by the discovery of
    'stone houses' on islands in the Dampier Archipelago

(*Murujuga*), which date from more than 8,000 years ago, and by changes in the subjects of its rock art.

51   Among the worst bush fires in recent Australian history were those on 7 February 2009 (Black Saturday) in Victoria, in which 173 people died and nearly 3,000 houses were destroyed. In south-east Australia, bush-fire incidence is projected to increase strongly by 2100, with longer fire seasons as a result of higher temperatures. Besides posing direct risks to livelihoods and property from burning, bush fires in Australia are also implicated in increased exposure of urban dwellers to atmospheric pollutants.

52   This and the previous quote come from Carnegie (1898: 17).

53   Much of this information comes from the excellent review of desert Aboriginal water management by Bayly (1999, *Journal of the Royal Society of Western Australia*). He includes a more graphic account of frog-water-drinking from the 1930s: 'My amazement reached a climax when he [an Aboriginal escort] seized the frog, placed the head-end in his mouth and squeezed its body. And while he squeezed he drank! It was not a mere sip, either; I should say the fluid he swallowed would have been sufficient to fill a teacup. As he drank, the old fellow looked at me out of the sides of his eyes in a very quizzical way; and when he had drained this most remarkable goblet to its last drop he smacked his lips afresh and exclaimed "Bullya Marra" ["Good! Good!"]' (p. 23). When the frog's water was all drunk, the rest of the frog was duly eaten.

54   Referring to the Aboriginal people of the Western Desert, Scott Cane found that 'almost every Aboriginal person I have spoken to can recall times when they were close to death – or members of their immediate family had died in attempts to find water' (p. 157 in Meeham and White 1990).

55   And most Aboriginal communities did live well for most of the time before 1788 in Australia. I note the comment of anthropologist Donald Thomson about the Pintupi (Bindibu) people of the central desert of Australia, apparently one of the harshest environments the country has to offer, that they 'have adapted themselves to that bitter environment so that

they laugh deeply and grow the fattest babies in the world'
(Thomson 1975: 4).

56 In February 2016, Severe Tropical Cyclone Winston – the
strongest hurricane ever recorded in the southern
hemisphere – tore a track through the islands of Fiji and
Tonga, leaving massive destruction in its wake. A friend told
me what happened in one of the affected villages in Fiji. The
people were gathered around a radio listening to official
warnings about the cyclone's path, on the basis of which they
decided *not* to evacuate the community to the caves inland. At
the time, meteorologists thought the cyclone was moving
away from the area where the village is located. But the older
people in the village thought otherwise because the birds were
flying unusually close to the ground, a traditional precursor of
an approaching hurricane. Science was favoured by the village
decision-makers and, when the cyclone was found to have
changed course, razing the village, it had a human impact that
might have been avoided had tradition been favoured over
science. I am grateful to Dr Lavinia Tiko for this story.

57 Modelling suggests that the ice-age population of Australia
may have fallen overall because of aridity (Williams, 2015,
Quaternary Science Reviews). It is certainly clear that people at
this time (14,000–28,000 years ago) abandoned vast tracts of
land where they had formerly lived because of the lack of
water, shifting to better watered refugia like the Murray-
Darling River Basin and south-west Tasmania. Over several
millennia (5,000–11,000 years ago) after the end of the last
ice age, Australian Aboriginal populations grew in every part
of the continent as both temperatures and rainfall increased.

58 An intriguing question is whether the Aboriginal groups that
reoccupied formerly abandoned areas of Australia knew that
their ancestors had lived there – and had kept that knowledge
alive during the millennia they spent in refugia – or whether
they merely saw the formerly unoccupied environments
becoming habitable and moved there. The most parsimonious
explanation – the type science invariably favours – is the latter,
but in the light of the likelihood that Aboriginal stories can
survive for 7,000 years or more (see Chapters 3 and 4), we
should perhaps not be too quick in dismissing the former.

Chapter 3: Australian Aboriginal Memories of Coastal Drowning

1   Australia's snake species are many and disproportionately
    venomous. There are about 140 species of land snake and
    many are deadly, even though they account for a mere 4–6
    deaths each year. An *Australian Geographic* article on the topic
    can be found at http://tinyurl.com/o6v45aa.

2   This argument hinges on the plausible idea that the Rainbow
    Serpent represents an amalgam of observations of 'snakes
    slithering away from drowning landscapes, rainbows
    overhead and strange "new" creatures such as pipefish washed
    ashore' (Tacon, 1996, *Archaeology in Oceania*, p. 117), a
    metaphor that helped people make sense of the rapid changes
    to the landscape they were witnessing.

3   At Delphi, the priestesses of the Oracle sat above cracks in the
    ground from which these gases (rich in CO_2-H_2S or ethylene
    or CH_4) rose, going into trances and purporting to predict
    future events. Geologists who examined the site found that
    no gases are escaping today, but that its structure shows that
    they once did, earthquakes periodically altering the
    circulation and storage (in shallow chambers) of gases
    produced from movements of hydrothermal fluids deeper
    within the Earth's crust. It is probable that the Delphic
    Oracle remained in place for the few hundred years that it
    took for a particular 'gas-exhaling chasm' to empty itself.
    This study was carried out by Luigi Piccardi who, together
    with Bruce Masse, deserves credit for helping make the topic
    of geomythology respectable among scientists, through a
    dedicated session (standing-room only!) at the 2004
    International Geological Congress and publication of a
    derivative volume (Piccardi and Masse 2007).

4   This is the main focus of my 2009 book, *Vanished Islands and
    Hidden Continents of the Pacific*, which reviews stories about
    'vanished islands' from many Pacific Island cultures, concluding
    that many represent memories of actual events. Some of the
    most plausible are the disappearances of the islands of
    Teonimenu and Vanua Mamata in the Solomon Islands and
    Vanuatu (South-west Pacific) respectively (Nunn 2009).

5   Loch Ness lies directly above the most seismically active
     part of the Great Glen Fault, one of the most active in
     Scotland. When there are earthquakes in the area, the
     surface of Loch Ness is often agitated, waves being created
     that may have been mistaken for the thrashing of a sea
     monster. This explanation is strengthened by noting that the
     earliest Latin descriptions of the appearance of the 'dragon'
     in the lake noted that it was accompanied 'with strong
     shaking' (*cum ingenti fremitu*) and disappeared 'shaking
     herself' (*tremefacta*).

6   See Ross (1986, *Oral Tradition*).

7   From p. iii of Dawson (1881).

8   All quotes in this paragraph are from Volume 1 of Matthew
     Flinders's journal, *A Voyage to Terra Australis* (Flinders 1814).
     He first entered Spencer Gulf on 9 March 1802 and left it 11
     days later. The night before, there was an eclipse of the
     moon and one wonders whether the Narungga associated
     this with the arrival of the *Investigator*, an event that would
     come to profoundly challenge their view of the world and
     their situation within it.

9   This story was sung in 1928 by a Wirangu woman named
     Susie from Denial Bay (Eyre Peninsula), and is recorded on
     p. 16 of Cooper (1955).

10   This account is from pp. 168–169 of Smith's 1930 book.
     There is reason in this instance to suspect that Smith
     plagiarised the work of David Unaipon, a tireless promulgator
     of Aboriginal culture (Krichauff 2011).

11   One detailed account was collected by anthropologist
     Charles Pearcy Mountford from the Adnyamathanha
     (Aboriginal) people living at the Nepabunna Mission in the
     Flinders Ranges (South Australia) in 1937. It states that
     Spencer Gulf 'was once a valley filled with a line of fresh-
     water lagoons, stretching northwards for a hundred miles or
     more. Each lagoon was the exclusive territory of a species of
     water bird. One lagoon belonged to the swans and the ducks,
     another to the grebes and the cormorants, still another to the
     water-hens, coots and reed-warblers. The trees belonged to
     the eagles, crows and parrots, while in the open country

between the lagoons lived emus, curlews and mallee fowls. Further out were the animals, the dingoes and many kangaroo-like creatures, and in the thick grass by the waters were the snakes, goannas and lizards' (Roberts and Mountford 1989: 18).

12   All quotes in this paragraph are from p. 18 of Roberts and Mountford (1989).

13   Native to Australia, willie wagtails (*Rhipidura leucophrys*) are birds that often appear in Aboriginal stories of different kinds, perhaps because they always seem so busy and sing so uniquely.

14   The quote in this paragraph comes from p. 172 of Smith (1930), while the idea that flooding of Spencer Gulf restored harmony between the competing groups of animals comes from Roberts and Mountford (1989), who probably derived their version of this Narungga story from a different source to that of Smith.

15   This and the previous quote are both from Volume 1 of Flinders (1814). 'Naive' fauna such as Flinders found on Kangaroo Island do indeed indicate a lengthy period of evolution without predators, as he surmised. Initial discovery by humans of many isolated oceanic islands also met with similar creatures, whose inability to recognise and avoid human predators invariably led to their extirpation or even extinction.

16   When Ronald Lampert started his PhD on the archaeology of Kangaroo Island in the 1970s, 'the problem had all the characteristics of a classic mystery story: a large offshore island without people, separated from the mainland nearly 10,000 years ago, yet with abundant evidence for a prehistoric population' (Lampert 1981: 1).

17   There are at least two extant accounts, collected independently, written down in the nineteenth century (Meyer 1846, Taplin 1873), and several more recent in date. The most comprehensive is that of Berndt (1940, *Oceania*), which was collected in the 1930s from the Jaralde people living in the lowermost part of the Murray River. The text in Chapter 3 includes elements from several versions of the Ngurunduri story.

18 The only direct indigenous version of this story to be written down, by David Unaipon in the 1920s, paints a different picture, in which Ngurunduri is portrayed as a good man whose affections were captured by two devious maidens he rescued from their imprisonment in a grass tree near Lake Albert. While Ngurunduri was temporarily absent, they caught and ate forbidden fish, for which he decided they should be punished. So they ran away from him (Unaipon 2001).

19 If this seems implausible, consider that Aboriginal women of that time had a different way of bathing from men (Berndt, 1940, *op cit*, footnote 36). Women often played 'water games' that involved hitting the water with the flats of their hands; 'she then very quickly uses her index finger, poking it in and out of the water she has divided, the result being a popping sound' (p. 179) that can sometimes be heard far away.

20 Quote from p. 181 of Berndt (1940, *op cit*). Other expressions of this key fact state that these events happened 'when the island was connected with the mainland [via] a strip of land' (Unaipon 2001: 131), or that 'Kangaroo Island ... was separated from the mainland only by a line of partly submerged boulders' (Roberts and Mountford 1989: 24), or that there was a shallow channel between the two, an isthmus in some accounts, across which a person could wade (Reed 1993, Parker 1959).

21 These words are from the Jaralde account given in the 1930s (Berndt 1940, *op cit*, p. 181). A story collected in the 1920s states that Ngurunduri began singing the Wind Song when his wives were halfway across. *'Pinkell lowar mia yound, Tee wee warr, La rund, Tolkamia a tren who cun, Tinkalla!* (Fall down from above, oh thou mighty Wind; swiftly run and display thy fleetness! Come thou down from the Northern sky, oh water of the deep! Come up in a mighty swell!)' (Unaipon 2001: 132).

22 From p. 57 of Taplin (1873).

23 From p. 181 of Berndt (1940, *op cit*).

24 From p. 182 of Berndt (1940, *op cit*).

25 This and all other quotes about MacDonnell Bay come from pp. 22–23 of the book by Smith (1880).

26 In addition to being used the seeds and roots of wattles
    (acacias) as food, the gum that many exude was used as a
    traditional medicine by Aboriginal Australians and was also a
    sweet treat for children.

27 The most comprehensive account of the geological history of
    Port Phillip Bay is that by Holdgate and others (2001,
    Australian Journal of Earth Sciences), while the issue of whether
    it dried up several millennia after the sea level had reached its
    present level is addressed by a similar group (2011, *Australian
    Journal of Earth Sciences*).

28 The quote is from p. 193 of Georgiana McCrae's biography
    (Niall 1994). It is likely that her main informant was
    Benbenjie, the 'premier huntsman and fisherman' of the
    Bunurong, who became her 'particular friend' and helped her
    compile a dictionary of the Bunurong language.

29 From p. 176 of McCrae (1934).

30 Both quotes from p. 12 of Hull (1859, *Report of the Select
    Committee of the Legislative Council on the Aborigines, 1858–9
    (Victoria)*.

31 Quote from p. 49 of Rogers (1957).

32 This may be a memory of a river flood ('rolling down') that
    drowned Port Phillip Bay, rather than inundation by the
    ocean or, more likely, a fusion of memories about change and
    causes that renders the extant story slightly ambiguous. This
    quote is from pp. 47–48 of Massola (1968).

33 Aboriginal Tasmanians were among the worst treated in the
    early decades of colonial Australia and were once thought to
    have died out with the death of Truganini in 1876, although
    this view was clearly wrong (Ryan 2012). In 1831 it was
    reported that Aboriginal people from Tasmania's east coast
    stated that 'this Island was settled by emigrants from a far
    country, that they came here on land, that the sea was
    subsequently formed' (Robinson 2008: 514, ML.A7085.2).

34 The quote is from pp. 267–268 of Fison and Howitt (1880).
    The finest *turndun* or bullroarers were made from cherry-tree
    wood and were used in men's initiation ceremonies in
    Aboriginal Australia, to which women and uninitiated
    children were not privy.

35 From p. 101 of Tench's (1793) *A Complete Account of the Settlement at Port Jackson*. In a footnote Tench stressed his point. 'The words which are quoted may be found in Mr Cook's first voyage, and form part of his description of Botany Bay. It has often fallen to my lot to traverse these fabled Plains; and many a bitter execration have I heard poured on those travellers, who could so faithlessly relate what they saw.'

36 The two quotes in this paragraph come from pp. 11 and 14 of the story of the Gymea Lily in the chapter by Bodkin and Andrews in the collection by Kuhn and Freeman (2012). This crimson-flowered plant, *Doryanthes excelsa*, is endemic to coastal areas around Sydney.

37 This story is part of an account by Bruce Howell in the 2016 volume of the *Sutherland Shire Historical Society Bulletin* entitled 'The Man They Called Mister (An Aboriginal Man living on Gunnamatta Bay in the 1920s)', which is based on conversations with James Cutbush in December 2015 that in turn recalled stories told him by his father Bill Cutbush, who spent his childhood in the area in the 1920s, befriending the man they called Mister.

38 Bailer shells (*Melo* sp.) are marine molluscs, often beautifully decorated, that can reach more than 40cm (16in) in length and have been commonly used in Australia and elsewhere for bailing water from canoes.

39 Quote from pp. 17–18 of Noonuccal (1990).

40 Bopple bopple trees are likely to have been either *Hicksbeachia pinnatifolia* or *Macadamia integrifolia*, both of which bear edible fruits/nuts, making them of particular interest to children.

41 The details in this paragraph are paraphrased from the account of Thomas Welsby (1967: vol. 2, 34), who probably collected this story himself.

42 This possibility is raised by O'Keeffe (1975, *Proceedings of the Royal Society of Queensland*), who also concedes that Cook's apparent observation of a single island 'may have been an illusion caused by the angle from which he viewed the place' (p. 85).

43 At the time of Cook's observations of Moreton and North
    Stradbroke Islands in 1770, the world was still in the grip of
    the Little Ice Age – most studies consider that it spanned
    about AD 1400–1800. During the Little Ice Age the sea level
    in this part of the world was lower than it is today during the
    Little Ice Age (Nunn 2007), so it is possible that the two
    islands were joined at this time.

44 This story is recounted by Dixon (1972: 29).

45 See Dixon (1980: 46).

46 Details from Gribble (1932), the quote from p. 57. Another
    version of this story (McConnel, 1935, *Art in Australia*),
    collected independently from the same area in the 1930s,
    notes that Goonyah (Ngúnya) had been using fish poison,
    probably from the leaves of *Derris trifoliata*, to catch fish: a
    practice often frowned upon in subsistence societies because
    it kills fish indiscriminately and can pass on the toxins they
    absorb to people who consume them.

47 For example, Dixon's Yidinjdji informant Dick Moses
    explained that Goonyah (Gunya) placed a sacred *woomera*
    (spear-thrower) in the prow of his boat to attempt to 'calm
    the waves', but to no avail. '*Bamaay ginuuy daguulji/banaang
    miwalnyunda gadangalnyuun/*; [The sea was coming in bringing
    with it] the three people and the canoe. The water was lifting
    up and bringing in [the canoe with the people in it] … *Balur
    ganaanggarr jarraal/banaagu wanggungalna/budiilna bana/*; [They]
    stood the curved *woomera* up in the prow [of the boat], to
    calm the water, to make it lie flat' (Dixon 1991: 92). In this
    story, we find echoes of the attempt by the English king
    Canute to command the waves to stop rising and, more
    pragmatically, an early account of people's increasing efforts
    to halt the encroachment of the sea onto the land.

48 From pp. 348–349 of McConnel (1930, *Oceania*). A related
    story comes from the Dingaal Aboriginal people, who
    occupy the Cape Flattery area about 200km (125 miles) north
    of Cairns, and have traditional title claim to Lizard Island
    (Dikaru), a high granite island just over 30km (19 miles)
    north-east of Cape Flattery and sheltered by the Great Barrier
    Reef to the east. Gordon Charlie, a Dingaal spokesman, 'says

that his direct ancestors once walked to the island from Cape
Flattery. When the sea levels rose and the Aboriginal people
could no longer walk there, they paddled their canoes from
island to island to reach it' (Falkiner and Oldfield 2000: 9).

49   Details in this section come from pp. 14–15 of Dixon (1977);
note that I have anglicised Yidin words for ease of reading.

50   During the last ice age, much of Australia was drier than it is
today, yet in contrast to other lake basins on the continent,
Lake Carpentaria is never known to have dried up completely
within the period 12,200–35,000 years ago. It was a large
lake, covering an average area of some 35,000km^2
(13,500mi^2), and containing 48km^3 (12mi^3) of water for most
of the time it existed. The beginning of its end came about
12,200 years ago when rising sea level overtopped the
Arafura Sill and seawater spilled into the lake. Its conversion
to full marine conditions was completed by about 10,500
years ago. There is a sole Aboriginal story about the
formation of the Gulf of Carpentaria, but its resemblance to
that of the kangaroo with the giant digging bone retold for
Spencer Gulf, to which it seems uniquely suited, suggests that
it may not be an authentic tradition for the Gulf of
Carpentaria (details in Reed 1965: 189–192).

51   You might think that given the comparative plethora of
drowning stories from elsewhere along the Australian coast,
there would be some among the peoples of the Torres Straits
Islands. Yet despite considerable anthropological and
linguistic research in these islands, no stories have been
formally reported, although Dixon alludes to a 1988
conference presentation in which Ephraim Bani, a Torres
Strait islander, spoke of 'ancient legends' that tell of people
'who actually walked as if on dry land, from the Australian
mainland to the Papuan coast' (Dixon, 1996, *Oceania*, p. 129).

52   The story tells that 'the two Djanggau Sisters, Daughters of
the Sun, came in their bark canoe with their Brother from
the mythical land of the dead, Bralgu, somewhere in the Gulf
of Carpentaria' (Berndt and Berndt 1994: 16). Another
Yolngu tradition about Bralgu [Island] is its link with the
planet Venus, known as Barnumbir and associated with

death. Two old women on Bralgu hold Venus on a long
string to ensure that it cannot escape (Bhathal, 2009, *Journal
and Proceedings of the Royal Society of New South Wales*).

53 This and the quote in the following sentence both come from
p. 20 of Dick Roughsey's autobiography, *Moon and Rainbow*
(1971).

54 The channel-cutting stories were collected by Paul Memmott
and analysed as part of his 1979 PhD. Ages of 5,000–5,500
years ago were assigned to them after consideration of the
history of postglacial sea-level changes in the Gulf of
Carpentaria. These ages are superseded by those presented at
the end of Chapter 4 in the present book.

55 It may seem self-evident to us that people would not
deliberately strand themselves on islands becoming cut off
from the mainland. Yet we should be wary of superimposing
our own views – what we would do in a particular
situation – on people with quite different worldviews
confronted by the same situation in the past. A good
analogue that I have been researching for some time is the
issue of how Pacific Island coastal communities should
respond to the sea-level rise they have been experiencing for
some decades, and which appears to be accelerating. The
'Western' approach is to understand the context of the
problem, evaluate scenarios for future sea-level rise, and
explain to coastal dwellers that they should move. But that
ignores the ways in which island people have long coped with
environmental adversity; it ignores their spiritual beliefs and,
for as long as they are given messages in foreign languages
(like English) that privilege foreign thinking, it seems likely
that they will continue to resist such exhortations. See also
my 17 May 2017 article in *The Conversation* at https://tinyurl.
com/khzm6nl.

56 From p. 194 of Josephine Flood's (2006) *The Original
Australians*.

57 Quote from p. 108 of Isaacs (1980).

58 I am struck by the similarity between the 'sandbars' of the
Yolngu Ancestresses and the sarns of Celtic cultures. Sarns –
literally highways – are linear sedimentary features that

extend from the land out across the sea floor, probably
ancient moraines or levees bulldozed into place by growing
glaciers, whose form and composition have been modified by
the sea following their submergence after the last ice age.

59 This story is recounted on p. 40 of Berndt and Berndt's
(1994) magisterial *The Speaking Land*.

60 This story was told originally by Peter Namiyadjad in the
Maung language (Berndt and Berndt 1994: 40). A more
recent version, in which the main actors are
anthropomorphised, was collected by Siri Veland (2013,
Global Environmental Change).

61 Quote from p. 25 of Corn (2005, *Journal of Australian Studies*).

62 Story told by Mangurug of the (northern) Gunwinggu
people, reported on p. 88 of Berndt and Berndt (1994).

63 In my book *Vanished Islands and Hidden Continents of the Pacific*
(Nunn 2009), I quote the examples of Teonimenu in the
Solomon Islands and Malveveng and Tolamp in Vanuatu as
places where the (abrupt) sinking of an island was
accompanied by giant waves that observers believed to be the
cause (rather than a consequence) of earthquake-induced
island submergence.

64 From pp. 88–89 in Berndt and Berndt (1994).

65 This is the assumption, made for the sake of analytical
completeness, by myself and Nick Reid in Table 1 of our
2016 paper in the *Australian Geographer*, freely available online
at http://tinyurl.com/hl8bdgt.

66 This quote is from p. 14 of Morris (2001). Details of the Tiwi
stories are from this source as well as from the account of
Sims's chapter in the (1978) collection by Hiatt.

67 From p. 165 of Sims (*op cit*).

68 Quotes from p. 10 in Morris (2001).

69 The language of the Tiwi islanders and the ways in which
their society is organised make them distinct from mainland
Aboriginal groups, differences that are attributable to the
isolation that began when the Tiwi Islands became separated
from the mainland by the rising sea level.

70 A parallel record of the isolation of the Tiwi Islands by
postglacial sea-level rise may be the 'extraordinary' level of

endemism shown by its rainforest ant fauna. Of the 34 species of ant found there, only nine are found anywhere else; the other 25 have been found only on these islands and perhaps evolved only after all land connections with the mainland became severed (Andersen, 2012, *Insectes Sociaux*).

71 Recollections of the effects of postglacial sea-level rise along other Tiwi coasts may include Myth 1 in the collection by Osborne (1974), which tells of a man stamping his foot repeatedly, causing the sea to rise over the land.

72 A readily accessible summary was published in *The Conversation* (online) on 6 April 2015.

73 This research was presented by Jo McDonald (2015, *Quaternary International*) and depended on the temporal sequencing of rock art in Murujuga. McDonald's continuing research in these islands uncovered evidence that their former Aboriginal inhabitants had built huts on stone platforms, something that many had previously thought Aboriginal Australians had never done.

74 None of these stories has been published. Most details here were supplied by Dr Katie Glaskin (University of Western Australia), who collected the majority of them as part of her 2002 PhD thesis; others come from testimony given by various Bardi and Jawi informants (Jimmy Ejai, Khaki Stumpagee and Aubrey Tigan) to a federal court case (Sampi vs State of Western Australia, 2005, FCA 777).

75 From p. 8 of the *Dictionary* (Moore 1884).

76 Much of the oldest evidence for human occupation of Rottnest is found within ancient soils (palaeosols) that have been buried by younger lithified sand dunes. These dunes contain the remains of land snails whose shells were able to be dated using aspartic acid racemisation assays. Dates on these shells of more than 50,000 years suggested to Patrick Hesp and others (1999, *Australian Archaeology*) that this was a minimum age for the human artefacts, although more recent work using optically stimulated luminescence dating suggests that the artefact-bearing layer is 10,000–17,000 years old (Ward, 2016, *Journal of the Royal Society of Western Australia*).

77 This account is from p. 341 of Mathews (1909, *Folklore*), and
    is part of a number of Aboriginal stories from the area that
    were collected by one Thomas Muir.

78 The story is found in Collet Barker's journals (Mulvaney and
    Green 1992: 361).

79 Quote from p. 157 of Maio and others (2014, *Palaeogeography
    Palaeoclimatology Palaeoecology*).

80 All the quotes in this section come from pp. 142–146 of Ethel
    Hassell's (1935, *Folklore*) collection of folktales from the
    Wheelman (Wiilman) tribe made before 1930.

81 Annual rainfall in the Nullarbor averages 150–250mm, but
    potential evaporation is closer to 2,000–3,000mm –
    unarguably a desert.

82 This is not strictly accurate. Between three and six million
    years ago, enough rain was falling on the Nullarbor for water
    to percolate into the subsurface limestone caves and form
    speleothems (stalactites and stalagmites). Pollen recovered
    from these suggests that gum trees (*Eucalyptus* sp.) and
    banksias (*Banksia* sp.) dominated a mesic forest that grew
    across the Nullarbor – at that time – in stark contrast to its
    present-day situation.

83 The Roe Plain(s) is part of the formerly more extensive
    coastal plain that existed during the lower sea level of the last
    ice age, but which is visible today only because it was uplifted
    shortly after it formed during Pliocene times, between 2.5
    and 5.5 million years ago. It comprises a veneer of
    calcarenites – sediments laid down in shallow ocean water –
    covering an erosional platform in the bedrock limestone that
    formed when the Pliocene cliffs of the Nullarbor were
    receding at an astonishingly rapid rate: perhaps 85km (53
    miles) in three million years.

84 Mallee eucalypt is the name given to at least three species.
    They have adapted to semi-arid conditions in Australia by
    developing long roots that extend laterally outwards,
    sometimes tens of metres, in search of water in every
    direction from the base of the tree. They also have roots that
    can penetrate downwards almost 30m (98ft). In 1928, in the
    desert north of the Nullarbor, one Archer Russell was being

guided by an Aboriginal man named Tuck who spotted 'a clump of big scrub-mallee'. Russell tells the story of what happened next. 'The trees I now noticed, had roots with sections growing alternately above and below the ground, and all the roots were long and twining. With Tuck wielding the shovel a root was soon exposed and torn from the ground. It was thirty feet (nine metres) long and no thicker than a man's wrist ... it is in reality an underground stem or rhizome. In each rhizome, which often contains a length of fifty feet (15 metres) or more, is enough water to sustain a man for a day' (Russell 1934: 100–102).

85 Quotes in this section are from pp. 89 and 91 of Scott Cane's (2002) comprehensive account of the Pila Nguru or Spinifex people who live in the Nullarbor.

86 Quote from p. 15 of Wright (1971).

87 Quotes from pp. 104–105 of Flinders (1814, vol. 1).

88 Quote from p. 401 of Berndt and Berndt (1996). Another version of this story – told by Mushabin (Bidjandjara), Harry Niyen (Antingari) and Marabidi (Ngalia-Andingari) – states that the brothers camped at Won-genya near Fowler's Bay and that, after the skin was pierced, 'all the water spread across the countryside and flowed down to the coast to become the Southern Ocean' (p. 44). Still another version – told by Sugar Billy Rindjana, Jimmy Moore and Win-gari (Andingari) and by Tommy Nedabi (Wiranggu-Kokatato) to Ronald Berndt in 1941 – adds the detail that the sea flood was prevented from spreading over the whole country by the 'action of various Bird Women' who gathered the roots of the *ngalda kurrajong* tree (probably *Brachychiton gregorii*), placed them along the coast to make a barrier and so 'restrained the oncoming waters', a response similar to that of the Wati Nyiinyii at the foot of the Nullarbor cliffs.

89 Assuming that, on average, a person acquired knowledge of these stories at the age of 15 and passed them on at the age of 35, this means that we count 'generations' for this purpose as lasting 20 years. Thus for a story to be passed on for 7,000 years (a minimum for these stories), it must be passed down through 350 generations.

Chapter 4: The Changing Ocean Surface

1   The earliest known remains of modern humans (*Homo sapiens*)
    are found within the Kibish Formation in the valley of the
    Omo River in southern Ethiopia. They date from about
    195,000 (± 5,000) years ago (McDougall and others, 2005,
    Nature). There is evidence for *Homo sapiens* living at Asfet on
    the Red Sea coast of Eritrea as early as 150,000 years ago,
    although the main period of occupation would probably have
    been during the subsequent interglacial, when this coastal
    region became better watered and altogether more attractive to
    these of our ancestors.

2   The Quaternary Period began 2.6 million years ago and is
    characterised by regular climate (glacial-interglacial)
    oscillations divided on the basis of oxygen–isotope values for
    ocean-floor sediments of various ages. There are 104 Marine
    Isotope Stages (MIS) within the Quaternary; each even-
    numbered stage marks a cool (glacial) period or ice age, and
    each odd-numbered stage marks a warm (interglacial) period.
    The Last Interglacial shown in Figure 4.1 is thus MIS 5, while
    the coldest part of the Last Glacial (the last ice age) is MIS 2.

3   An argument along these lines was laid out to explain the
    start of cross-ocean discovery and occupation of oceanic
    islands in the Western Pacific Ocean in my book
    Environmental Change in the Pacific Basin (1999).

4   Information comes from the study by Scott and others (2014,
    Antiquity), who argue that the bone accumulations at La Cotte
    represent the final resting places of the megafauna that were
    extirpated here because they were unable to withstand the
    onset of cold conditions within the Last Glacial. There is a
    range of ages for the human and faunal presence at La Cotte,
    from 40,000 to 238,000 years ago, but another study quotes
    the more recent terminal age of 25,700 ± 3,000 years ago that
    is noted in the text (Bates, 2013, *Journal of Quaternary Science*).

5   The most comprehensive study of Doggerland is the book by
    Gaffney and others (2009).

6   Such remains began to be recovered in earnest after a Dutch
    fishing innovation called beam trawling, which involved
    dragging weighted nets along the ocean floor, was developed;

'sometimes an enormous tusk would spill out and clatter onto the deck, or the remains of an aurochs, woolly rhino, or other extinct beast. The fishermen were disturbed by these hints that things were not always as they are. What they could not explain, they threw back into the sea' (Laura Spinney, *National Geographic Magazine*, December 2012).

7   The sabre-toothed cat (*Homotherium latidens*) lived in Europe, Asia and North America. In size and musculature, it more closely resembled modern lions than modern tigers. Most of the details in this section come from the book by Mol and others (2008).

8   In short, the wetter times of the last ice age appear to coincide with periods of expansion of areas occupied by modern humans (*Homo sapiens*), the proposed connection being that more rain would have made drier places more attractive to our ancestors because of the increased livelihood possibilities associated with the establishment of vegetation diversity. Using analyses of speleothems (dripstone deposits like stalagmites and stalactites), one research project reconstructed rainfall in North Africa during the last ice age and found just this for human expansion within and beyond this region. With remarkable precision, it proved possible to link a wet phase (50,500–52,500 years ago) to the movement of modern humans out of North Africa into the Middle East, where they first encountered – and interbred with – Neanderthal humans (Hoffmann, 2016, *Scientific Reports*).

9   Megafauna in most parts of the world became extinct approximately simultaneously in the latest Pleistocene, typically by 10,000–13,000 years ago. In some places, the weight of evidence for megafaunal extinction favours predation by increasing numbers of increasingly well-armed and cooperating humans, while elsewhere deglacial climate changes are implicated. Explanations involving both rapid climate changes and human predation have also been mooted for some places. See also Chapter 6.

10 One of the pioneers in the use of 'island dipsticks' for measuring postglacial sea-level change was one of my heroes,

the late Art Bloom of Cornell University, in many of whose
footsteps around the Pacific Islands I have followed.

11   It is unclear whether most of the meltwater from the collapse
     of this ice saddle entered the Pacific or Atlantic Oceans. The
     former perhaps received the larger part because the
     thermohaline circulation, which is most vulnerable to
     freshwater inputs in the North Atlantic, did not shut down
     during MWP-1A.

12   The argument is neatly articulated in Barber and others
     (1999, *Nature*).

13   An earlier study suggested that the instantaneous rise of
     the sea level in the Mississippi Delta may have been as much
     as 1.2m (nearly 4ft), although this probably includes a
     background sea-level rise component as well as the effects
     of the outburst flood.

14   There have been serious suggestions from sensible scientists
     that future sea-level rise might cause the West Antarctic ice
     sheet to collapse catastrophically, raising sea levels an average
     of 3.3m (11ft) along the world's coasts, albeit with significant
     regional variations (Bamber, 2009, *Science*). Similar
     suggestions have been made for the Greenland ice sheet
     (Shannon, 2013, *Proceedings of the National Academy of Sciences
     of the United States of America*).

15   The origin of rice agriculture, if indeed there was a single
     place of origin, is disputed and may be becoming an issue of
     national pride rather than detached empiricism. That said,
     most evidence points to the Yangtze as an important place of
     early rice domestication, although there is increasing
     evidence that it may not have been the sole place of origin.

16   The comparatively 'late' development of agriculture in the
     Vistula Delta has long puzzled archaeologists. In addition to
     the effects of an oscillating sea level, which may have
     frustrated attempts to establish enduring crop production and
     animal husbandry here, it may simply be that the collection
     and processing of amber proved more profitable for many of
     the Roman-era Pomeranians living here. In addition to
     bedrock deposits of amber suited to mining, the shores of the
     Gulf of Gdańsk are as famed today for the deposition of

wave-borne amber (often in nuggets) as they were in Roman times. From here, the amber was exported along an 'amber road' to Mediterranean markets, where it was in great demand for jewellery and for medicinal purposes.

17 The model for the Pacific Islands that links climate change and sea-level fall to an enduring food crisis about AD 1300, and a subsequent abandonment of coastal settlement in favour of hillforts, is described in considerable detail in my book *Climate, Environment and Society in the Pacific Islands during the Last Millennium* (Nunn 2007).

18 My most recent research into Fiji's hillforts, conducted in collaboration with the Fiji Museum, has been in the Bua area of Vanua Levu Island. A popular account of this project was published online in SAPIENS on 15 June 2016 and is downloadable from http://tinyurl.com/z9st6eb.

19 Quotation comes from p. 55 of Henry Britton's *Fiji in 1870*, a collection of articles he wrote for the *Argus* newspaper in Melbourne (1870).

20 The most comprehensive and accurate assessment of current and recent rates of sea-level change is found in the 5th Assessment Report of the Intergovernmental Panel on Climate Change (IPCC). The science was assessed by Working Group 1. The Summary for Policymakers is an accessible summary for non-specialists – see ipcc.ch.

21 Closure of the 'sea-level budget' for 1993–2010 was a major achievement of the 'Sea Level Change' chapter in the 5th assessment report of the IPCC. Until this point, there was uncertainty about exactly which cause/s of sea-level rise were most important. Settling the issue has implications not only for the understanding of past sea-level changes and the modelling of future ones, but also for strategies to mitigate the effects of future climate change.

22 The original area of the contiguous land mass of Sahul (comprising modern Australia, Papua New Guinea and Irian Jaya, or Indonesian Papua) during the Last Glacial Maximum about 20,000 years ago (when the sea level was about 120m/400ft lower than it is today) was $11,021,024km^2$ ($4,255,241mi^2$). The modern land area totals $8,473,836km^2$ ($3,271,766mi^2$), so

2,547,188km² (983,475mi²), or 23 per cent, was inundated when the sea level rose after the end of the last ice age.

23   This authoritative compilation of past sea level around Australia was developed by Stephen Lewis and others and published in 2012 in the journal *Quaternary Science Reviews*.

24   Delicate bivalves (like *Donax* and *Paphies* spp.) have been used in Australian sediment cores to identify former shorelines. Where these shells are found unbroken rather than pulverised, this indicates that they had hardly been moved from the coast they formed.

25   Six sediment cores collected from the sea floor between the Arafura Sea and Gulf of Carpentaria show that the sill separating the two was covered by seawater most recently about 9,700 years ago (Chivas, 2001, *Quaternary International*).

26   David Hopley's pace-setting 1982 book on the *Geomorphology of the Great Barrier Reef* made the case for coral-reef coring as a precise way of reconstructing past sea levels with an almost unmatched degree of precision. This book has more recently appeared in an updated version (Hopley, Smithers and Parnell 2007).

27   See Dixon (1980: 46).

Chapter 5: Other Oral Archives of Ancient Coastal Drowning

1   I am grateful to Marjorie Le Berre for information in these two paragraphs, which is based on conversations we had in Nantes in November 2016 and is used with her permission.

2   The city of Ys is sometimes placed not in the Baie de Douarnenez, but in the Baie des Trépassés, or in the Baie d'Audierné (Guyot 1979), all along the west-facing coast of Brittany.

3   English translation on p. xiii of the book by Guyot (1979), from Bertrand d'Argentre's (1588) *L'Histoire de Bretaigne* (The History of Brittany).

4   '*La ville d'Is fut submergé et presque tous ses habitants périrent*': French text from p. 22 of Sébillot (1899).

5   '*Les pêcheurs de Cancale dissent que quand la mer est belle et claire, on voit entre le Mont Saint-Michel et les îles Chausey de debris de*

murailles. Ce sont les restes d'une ville disparue': French text from
p. 23 of Sébillot (1899).

6　Quotes from p. 336 of Peacock (1865, *Proceedings of the Royal
Geographical Society of London*), who also concluded that the
monastery at Menden is now under 8m (26ft) of ocean, that
at Mandan (west of St Pair) now under 10m (33ft), that
named after St Moack now under 13–14m (43–46ft), and
Taurac (or Caurac) now submerged to a depth of 16m (52ft).

7　St Guénolé (or Guignolé) is often remembered in English-
language texts as St Winwaloe; details of his life were
summarised by Doble (1962).

8　Nowhere is this perhaps more obvious than on the delta of
the Brahmaputra-Ganges-Meghna river, occupied mostly by
several tens of millions of Bangladeshis, which has over
recent decades experienced storm surges that have reached
progressively further inland as a result of these being imposed
on sea-level rise. As the sea level rises in the next few
decades, so storm surges will reach even further inland.

9　*'Au temps jadis la Manche n'était pas si grande que maintenant; l'on
pouvait aller à Jersey sans rencontrer d'autre obstacle qu'un ruisseau
qui n'était pas trés large'* (Sébillot 1899: 23). In addition, a
contemporary story about 'the conjunction of Jersey to
Normandy' was alluded to by Poingdestre (1889: 75).

10　From *Gazette de l'Ile de Jersey*, 28 April 1787, quoted by
Peacock (*op cit*, p. 329), who identifies numerous other
instances of submerged forests in the Channel Islands.

11　Reading Alan Seeger's eponymous 1916 poem, written at a
time when the ancient tradition was probably better known
than it is today, Lyonesse sounds as though it might have
been Ys – as indeed it may:

> *In Lyonesse was beauty enough, men say:*
> *Long Summer loaded the orchards to excess,*
> *And fertile lowlands lengthening far away,*
> *In Lyonesse.*

> *Came a term to that land's old favouredness:*
> *Past the sea-walls, crumbled in thundering spray,*
> *Rolled the green waves, ravening, merciless.*

> *Through bearded boughs immobile in cool decay,*
> *Where sea-bloom covers corroding palaces,*
> *The mermaid glides with a curious glance to-day,*
> *In Lyonesse.*

12   From an 1854 translation of *The Chronicle of Florence of Worcester*, quoted in Hunt (*op cit*) at www.sacred-texts.com/ neu/eng/prwe/prwe085.htm, accessed in April 2009.

13   Quoted by Hunt (*op cit*). Other authoritative sources that name Lyonesse and place it off the coast of Land's End include *Britannia* (Camden 1590) and the 1602 *Survey of Cornwall* (Carew 1723). Camden was an antiquarian who collected folk tales. He concluded that Lyonesse (the City of Lions) was located off Land's End where the Seven Stones reef now lies. Camden also reported that sailors could hear the bells of Lyonesse ringing when they crossed the area during heavy seas.

14   There are examples from the South-west Pacific island groups of the Solomon Islands and Vanuatu in which islands have reportedly vanished as a result of large-wave impact but which must also have been affected, either during the earthquake that produced these (tsunami) waves or a short time after it, by rapid subsidence. It was the latter that caused the islands to disappear. Examples include Teonimanu and the Pororourouhu group in the Solomon Islands, and islands like Malveveng and Tolamp in Vanuatu. These and other examples are in my book *Vanished Islands and Hidden Continents of the Pacific* (Nunn 2009), available as an audio book at https://tinyurl.com/lbrrgfa.

15   A contemporary account notes that to the astonishment of onlookers on the south coast of Cornwall, 'the Sea rose near six feet [1.8m], coming in from the South-East extremely rapid; then it ebbed away with the same rapidity to the Westward for about ten minutes; till it was near six feet lower than before; it then returned again, and fell again … the first and second fluxes and refluxes were not so violent as the third and fourth, at which time the Sea was as rapid as that of a mill-stream descending to an under-shot wheel, and the rebounds of the Sea continued in their full fury for fully two

hours' (Borlase 1758: 53–54). This happened at a distance of around 1,500km (932 miles) from the epicentre. The Reverend Borlase described a more locally centred earthquake on 15 July 1757, during which 'on the island of St. Mary, Scilly, the shock was violent'.

16 The Roman count is found in Strabo's *Geography* (AD 17?), at a time when the Scilly Isles were called the Cassiterides.

17 Causeways and house foundations have also been found underwater within the Scilly archipelago (Ashbee 1974).

18 The story is immortalised in the Welsh-language poem *Clychau Cantre'r Gwaelod* (The Bells of Cantre'r Gwaelod) by J. J. Williams. The extract quoted here was translated by Dyfed Lloyd Evans and accessed at www.celtnet.org.uk in October 2007.

> *O dan y môr a'i donnau*
> *Mae llawer dinas dlos*
> *Fu'n gwrando ar y clychau*
> *Yn canu gyda'r mos.*
> *Trwy ofer esgeulustod*
> *Y gwiliwr ar y tŵr*
> *Aeth clychau Cantre'r Gwaelod*
> *O'r golwg dan y dŵr.*

> Beneath the wave-swept ocean
> Are many pretty towns
> That hearkened to the bell-rings
> Set pealing through the night
> Through negligent abandon
> By a watcher on the wall
> The bells of Cantre'r Gwaelod
> Submerged beneath the waves.

19 A much larger area of more than 6,200km^2 (2,400mi^2) is envisaged in the analysis of Flemming (1972), who was consequently sceptical that such a land ever existed. The name Cantre'r Gwaelod means 'the bottom cantref', a *cantref* being an area of perhaps 466km^2 (about 180mi^2 – Flemming 1972).

20 All quotes paraphrased from accounts quoted by Doan (1981, *Folklore*).

21 The quotes in this sentence and the next come from p. 81 of
    Doan (*op cit*), who also quotes authoritative sources for the
    age determination.

22 'It is to representatives drawn from among the famous
    legendary heroes of the sixth century, the period assigned to
    the beginning of their national traditions, that medieval
    cyfarwyddiaid [storytellers] attached the legends of the great
    inundations' (from p. 241 in the chapter by Rachel Bromwich
    in the collection by Fox and Dickens (1950)).

23 Quoted by Bromwich (*op cit*, p. 229) from a 1917 account by
    Richard Fenton.

24 This information is found on p. 334 of Wilson's (1870) *The
    Imperial Gazetteer of England and Wales*. The idea that human-
    made structures are visible at Caer Arianrhod was roundly
    criticised by North (1957) and others, yet they do not address
    the more subtle point about whether or not the memory of
    such a place may be authentic ... even if its currently
    favoured physical manifestation may not.

25 A submerged forest in the Dovey Estuary was dated to
    4,700–6,000 years ago when the sea level here may have been
    2–4m (6½–13ft) lower.

26 Another estimate by Kurt Lambeck suggests that there was a
    wide land bridge between Ireland and Wales from 13,000 to
    20,000 years ago, but it is not explained when it may have
    become impassable. Yet it is noted that about 14,000 years
    ago, because of the huge volumes of meltwater pouring into
    the Irish Sea, the shrinking land bridge would have been 'a
    swampy and inhospitable region at best' (1995, *Journal of the
    Geological Society of London*, p. 443).

27 The quotes come from p. 30 of a recent translation of the
    Mabinogion (Jones and Jones 2001). Renowned Anglo-Saxon
    scholar Rachel Bromwich commented that 'this passage may
    represent the eleventh-century tradition about the submerged
    lands' between Wales and Ireland and is a story that appears
    'quite independent' of that about Cantre'r Gwaelod (*op cit*,
    p. 228).

28 As an aside, it is fascinating to consider the possibility that an
    oral tradition like that of Brân walking from Wales to Ireland

may have seen him posthumously changed to a giant, for how else might anyone be persuaded to believe he had crossed an ocean on foot? It may be that an implausible tradition, preserved for millennia, led to the belief that giants once existed. And of course, once that particular fantastic door had been opened, any number of similar traditions became game for the involvement of giants.

29   These calculations are based on modelling of postglacial sea-level change by Lambeck (*op cit*). Cornwall is currently sinking at an estimated 0.6mm each year; assuming this rate applied to the last 10,000 years or so (which it probably did not), then we can ascribe some 6m (20ft) of submergence to it, which still requires the sea level to have been 60–70m (200–230ft) lower for these Lyonesse stories to be based on observation of a land bridge between the Scilly Isles and Cornwall.

30   In AD 1250, Dunwich housed more than 5,000 people living in at least 800 taxable dwellings, spread out over some 330ha (815 acres) (Sear, 2011, *International Journal of Nautical Archaeology*).

31   This effect can be explained by Global Isostatic Adjustment (GIA), which pictures the Earth's crust as continuing to respond to the removal of terrestrial ice loads during the last ice age. Areas that were thickly ice covered are rising because of the 'flow' of crustal (lithospheric) material towards them from areas that were not ice covered, or at least not so thickly. The zone of maximum subsidence in north-west Europe crosses the southern North Sea and can be considered a significant contributor to land sinking along the East Anglian coast.

32   Not to be confused with the Herakleion on the Mediterranean island of Crete.

33   These figures are those proposed by Stanley (2007). Founded in the sixth century BC, Herakleion thrived for some 600 years; Eastern Canopus was probably established a little later and remained visible until the mid-eighth century AD.

34   The most complete account of the Canopic distributary and the rediscovery of its Greek port cities is by Stanley (2007).

A comparable explanation refers to the 'disappearance' of the
city of Vineta in the Oder River Delta on the Baltic coast of
modern Poland or Germany.

35 A contemporary account of the disaster by Pausanias
described how 'the sea flooded a great part of the land and
encircled the whole of Helike. Moreover, the flood from the
sea so covered the sacred grove of Poseidon that only the tops
of the trees remained visible' (quoted in English in the
chapter by Soter, p. 41 in the book by Iain Stewart and
Claudio Vita-Finzi, 1998).

36 This remarkable piece of geological sleuthing was carried out
by Soter and Katsonopoulou (2011, *Geoarchaeology*).

37 Movements of extremely hot water and steam within the
Phlegrean Fields caldera, where these ports were located,
account for the sometimes frenetic movements of the Earth's
surface here. Hot water or gas, superheated by its proximity to
liquid rock below, forces its way upwards through the Earth's
crust, often pushing into chambers just below the surface and
causing them to expand, sometimes only temporarily, and the
ground surface to rise. Once these chambers empty, the
surface can fall. Alternate rise and fall of the ground surface
makes the geography of the Phlegrean Fields region one of
the most changeable in the world. In 1982–1984, an increase
in such 'bradyseismic' effects rendered the city centre unsafe
and raised the adjacent sea floor by some 2m (6½ft). Somewhat
ominously, the dominant cause of ground-surface changes in
the region changed in about 2012 from hydrothermal to
magmatic, that is from the subterranean movement of hot
liquids to the emplacement of a lava reservoir, as shallow as
3,000m (9,850ft), below the streets half a kilometre (⅓ of a
mile) from the seafront in Pozzuoli.

38 Most of the information about the seismotectonic history of
the Rann of Kachchh comes from the work of Bilham (in a
chapter in the book by Stewart and Vita-Finzi, cited above),
who painstakingly analysed all the qualitative accounts to
produce a plausible quantitative scenario that can be used to
inform future earthquake planning in this region. Reduced
crustal stress in the Kachchh area probably means that

another big earthquake here is unlikely in the foreseeable future, although accumulated stress may have been transferred along regional fractures towards Karachi, Pakistan.

39 Cited in Bilham (*op cit*, p. 5).

40 Quotes in this paragraph come from the Griffith translation of Hymn 87 (LXXXVII) accessed online in February 2017 at www.sanskritweb.net/rigveda/griffith.pdf.

41 The most comprehensive account of Dwaraka is the book by Rao (1999).

42 This information is found in the *Harivamsa*, an appendix to the *Mahabharata*. A *yojana* is considered to have been a measure of distance, perhaps 12–15km (7–9 miles), not area, so it is not straightforward to estimate how much land was reclaimed at Dwaraka. If we assume the extent of coastal protection structures (bunds and sea walls) to be 12km, then the area reclaimed may have been 5–6km^2 (about 2mi^2).

43 Never underestimate people's attachment to place, especially when this is imbued with considerable investment of money and effort. After the devastating impact of Hurricane Katrina on New Orleans, USA, in 2005, there were calls for the most vulnerable parts of this iconic city – those as much as 2m (6½ft) below sea level – to be abandoned and their occupants to be compensated and relocated elsewhere. It did not happen.

44 From the *Mausala Parva* (Book of Clubs), the sixteenth book of the *Mahabharata*.

45 The key details in stories transmitted orally often become compressed, an effect dubbed 'memory crunch' (Barber and Barber 2004).

46 Thermoluminescence (TL) dates were reported by Vora and colleagues (2002, *Current Science*). When applied to ceramics, the TL technique can measure the time elapsed since firing, a process that resets the TL clock to zero. By heating crystalline samples (as in bits of pottery), light is produced – the thermoluminescence – and the amount of this light is proportional to the radiation dose that has accumulated over time, something that can be converted to age.

47 Based on the work of R. N. Iyengar (2005, *Journal of the
    Geological Society of India*).
48 The Harappan cultural tradition appeared about 5,200 years
    ago (3250 BC) along the Indus River Valley, and spread,
    based on a system of complex urban centres sustained by
    agricultural production. The Sausashtra Peninsula, where
    Dwaraka was located, was on the margins of Harappan
    culture. It is possible that the Rann of Kachchh, today one of
    the least hospitable parts of India, was awash and navigable
    during Harappan times, something that would have
    transformed the productive and exchange potential of
    this area.
49 The story is told in Book Six (*Yuddha Khanda*) of the
    Ramayana. In the prose rendering of Sarga (Chapter) 22,
    Rama spoke harshly to the ocean – 'O, ocean! I will make
    you dry up now along with your nethermost subterranean
    region. A vast stretch of sand will appear, when your water
    gets consumed by my arrows; when you get dried up and the
    creatures inhabiting you get destroyed by me. By a gush of
    arrows released by my bow, our monkeys can proceed to the
    other shore even by foot. O Sea, the abode of demons! You
    are not able to recognize my valour or prowess through your
    intelligence. You will indeed get repentance at my hands'
    (from www.valmikiramayan.net/yuddha/sarga22/
    yuddha_22_prose.htm, accessed in February 2017).
50 These dates are reviewed in Krishnaswamy and Nandan
    (2005, *Journal of the Geological Society of India*), who reconciled
    geological evidence for a period of unusually cold winters in
    India with the description of these in the *Ramayana*.
51 The story of the failed bridge is found in the seventh-century
    poem *Setubandha*.
52 We cannot discount the possibility that tectonics – land
    movements – also had a role in the disappearances (and
    appearances) of Ramasetu. Uplifted coral reefs, some 4,000
    years old, are found on the Indian side of Ramasetu.
53 A likely chronology for the Pallava Era is from about AD 300,
    when Sivaskandavarman ascended the throne, until AD 869,
    when Nandivarman III died.

54 This information comes from Sundaresh and others (2004, *Current Science*), who also quote an eighth-century Tamil text describing Mahabalipuram as a place 'where the ships rode at anchor bent to the point of breaking, laden as they were with wealth, big-trunked elephants and gems of nine varieties in heaps' (p. 1231).

55 From an 1869 account quoted by Sundaresh and others (*op cit*, p. 1232).

56 Research reported by Rajani and Kasturirangan in the 2013 issue of the *Journal of the Indian Society of Remote Sensing*.

57 The Bay of Bengal, which is bordered by the east coast of India (where Mahabalipuram is located), is a structural feature that is experiencing subsidence of as much as 3mm per year resulting from regional tectonic movements in addition to the effects of both sediment loading of the ocean floor and sea-level rise. The Bay also experiences instances of rapid subsidence associated with earthquakes that have contributed to the progressive submergence of the Mahabalipuram area over the past few millennia.

58 To the south of Mahabalipuram on India's east coast, there are also stories about submerged places at Poompuhar and Tranquebar. Older perhaps are stories about a submerged continent, Kumari Kandam, off the southern coast of India, which may have influenced European stories about a supposed 'lost continent' named Lemuria in the Indian Ocean (Ramaswamy 2004).

Chapter 6: What Else Might We Not Realise We Remember?

1 The greenhorn was the author, who published a detailed analysis of the Kadavu volcano myths (1999, *Domodomo* [Fiji Museum]). Years later, the cutting of a new road around the base of Nabukelevu revealed volcanic scoria lying above soils in which potsherds were found, unequivocal evidence for eruptive activity here since the human settlement of Kadavu Island. Kadavu is part of a volcanic island arc, similar to many in the Western Pacific,

formed by volcanic activity linked to the underthrusting of one crustal plate beneath an adjoining one. The lower plate is forced into the Earth's interior where it melts, the liquid rock finding its way to the surface above – in places like Kadavu.

2   An absorbing study of the 'times of darkness' in highland New Guinea is by Russell Blong (1982), who employed mineralogical analyses of ash to determine which eruptions from which volcanoes had affected particular places – and might therefore do so in the future.

3   The quote comes from William Dampier's (1729) *A Continuation of a Voyage to New Holland*, accessed online through Project Gutenberg in February 2017; the original spellings are quoted on p. 1 of *Fire Mountains of the Islands* (Johnson 2013).

4   Lines of intraplate (not plate-boundary) islands in the world's ocean basins are explained as a result of an oceanic plate (piece of Earth's crust) moving over a fixed mantle plume, known as a hotspot. As long as the plume keeps leaking liquid rock, a succession of volcanoes will form, the only active one(s) being that closest to the hotspot (Nunn 1994). The best-studied example is the Hawaii-Emperor island-seamount chain that extends nearly 6,000km (3,750 miles) across the North Pacific and has been active for at least 80 million years. The youngest above-sea volcano in this chain is the Big Island (Hawai'i), which is nearing the end of its active life. Yet lurking 1.2km (¾ mile) beneath the ocean surface, 35km (22 miles) to its south-east, lies the active and growing volcano of Lo'ihi, destined one day to spectacularly break the ocean surface and grow into a massive structure like the Big Island is today.

5   There has been a lot of research on this topic. What has recently emerged as a more compelling explanation of NVP volcanism, supported by teleseismic tomography of the solid earth beneath the area, is edge-driven convection (EDC). This requires abrupt changes in crustal (lithospheric) thickness, expressed along its lower boundary by steps within each of which liquid rock is circulated by convection. Below

the NVP there is a step within which liquid rock is circulating, an anomaly in an Australian context, but likely to be responsible for NVP volcanism within the last five million years.

6 Many evangelical Christian groups interpret the time periods specified in the Bible literally, believing that the Earth since its divine creation is only a few millennia old. To justify this interpretation, it is then deemed necessary to discredit radiometric dating, something that a few authors have gone to tortuous lengths to attempt. A fine review of creationism and its historical antecedents is the book by Arthur McCalla (2013).

7 A billion here means 1,000,000,000. For some of the oldest rocks on Earth, other techniques involving even slower rates of isotope decay are used, including the awesome rubidium-strontium method in which ^{87}Rb decays to ^{87}Sr with a half-life of 50 billion years.

8 Tennateona was so cruel and savage that some people, to save themselves from him, 'laid themselves on ant heaps and let the ants cover their bodies as if dead' (Smith 1880: 24).

9 The earliest written version of this story (much was misquoted subsequently) is found on pp. 14–15 of the book by Christina Smith, a missionary who lived in the Mt Gambier area from 1854 until her death in 1893. She learnt to speak the local Bungandidj language and recorded numerous stories of the Buandig (Booandik) people (Smith 1880).

10 Dating of materials below (and therefore predating) the youngest lava flows at Mt Gambier show a maximum age for these of perhaps 4,300 years ago. A similar date of 4,930 years ago was obtained from Mt Schank.

11 The quotation is from p. 102 of Dawson (1881); other stories are reported herein and another about the eruption of Mt Buninyong by Howitt (1904).

12 A good example of the arbitrariness of history is that Mt Eccles was originally Mt Eeles, named for a prominent war veteran, but a draughting error in the 1850s rendered it 'Eccles', which name has been used since!

13   Depth-age calculation from Figure 4.3 in Chapter 4 shows
      that the sea level around Australia was 33m (108ft) below its
      present level between 10,150 and 10,750 years ago. A likely
      age for the Tyrendarra flow is around 36,000 years ago.
14   The name Gerinyelam, rendered more recently as
      Derrinallum, for Mt Elephant is recorded on p. 179 of Smyth
      (1878). The dates for the most recent eruptions of Mt
      Elephant are reported in an anonymous article but appear to
      be estimates.
15   Information from www.kanawinkageopark.org.au, accessed
      in March 2017.
16   For example, a 2017 study by Ben Cohen and others stated
      that 'the cause of the young lava field volcanoes in
      Queensland is not well known ... no single model satisfies all
      observations' (Quaternary Geochronology, p. 80).
17   Quote from p. 90 in Cohen's work (see previous note).
18   This date is actually a range from 5,000 to 9,000 years ago;
      the median age is 7,000 years ago. The unusually large error
      (± 2,000 years) is because a comparatively new dating
      technique ($^{40}Ar/^{39}Ar$) was employed; future refinements are
      likely to see error margins reduced.
19   I am not sure what to make of the Aboriginal story, recorded
      about 1917 'halfway up the Tully River', 100km (60 miles)
      north-east of Kinrara, which 'related how long ago ... fire
      and flames had erupted suddenly from the rocks and a rain of
      stones had fallen in the surroundings' (Mjöberg 1918: 141).
      The explanation given was that this had been the work of an
      evil spirit seeking revenge; it was noted that at that time,
      local people 'avoided getting too close to the place of the
      eruption'. The problem is that there is no known youthful
      volcano in this area, so it is possible this story came here from
      somewhere else.
20   This Mamu/Ngajan version of the story was told by George
      Watson in the Mamu dialect of Dyirbal. It is summarised on
      pp. 153–154 of linguist Robert Dixon's entertaining memoir
      (1984). An alternative version, told by Dick Moses in the
      coastal dialect of Yidiny, is written in both the original and
      translation in another of Dixon's books (Text 3 in Dixon 1991).

21   The Fujisan legend quoted comes from the landmark
     compilation by Dorothy Vitaliano (1973), which also contains
     examples from other parts of the world.

22   For a popular yet scientifically rigorous account of Krakatau,
     I recommend the book by Simon Winchester (2003).
     Indonesia was also the site of the even larger Toba
     supereruption about 75,000 years ago (on Sumatra Island),
     which produced some 2,800km^3 (670mi^3) of material; in
     1883 Krakatau produced about 12. By pouring so much
     airborne material into the Earth's atmosphere, the Toba event
     may have caused a 'volcanic winter' that blotted out the sun,
     causing plant growth to diminish to a point where humanity
     became imperilled; people everywhere starved to death. The
     Toba supereruption was once believed to have been
     responsible for a bottleneck in modern human evolution,
     although subsequent research suggests that this view might be
     overstated; people in India appear to have coped, and those in
     East Africa were apparently unaffected.

23   The date of 16,000 years ago for the draining of Lake
     Bandung near Tangkuban Perahu comes from research by
     Dam and colleagues (1996, *Journal of Southeast Asian Earth
     Sciences*). I have woven together several versions of the story
     of Sangkuriang and Dayang Sumbi, the principal published
     version being that in Vitaliano (1973).

24   This story is found in the memoir by Charachidzé (1986).

25   Using the potassium–argon (K-Ar) technique, lavas from
     Kazbek and Elbrus and numerous other Caucasian volcanoes
     have been dated (Lebedev, 2014, *Journal of Volcanology and
     Seismology*). In about AD 320, an eruption of Elbrus caused
     avalanches that buried ancient soils (palaeosols), which have
     been dated using radiocarbon to determine the age of the
     eruption (1,780 ± 70 years ago). The first indication that
     ice-capped Elbrus might be about to spring back to life was
     when scientists noted moss growing above cracks in the ice
     surface, suggesting that heat from beneath the ice was
     escaping to the surface along these cracks. Subsequent
     investigations confirm that the mountain's top is hotter than
     would be expected – it sits more than 5,600m (18,370ft)

above sea level – perhaps because of the filling of a shallow subterranean magma chamber (see https://sputniknews.com/voiceofrussia/2010/08/02/14236033.html, accessed in February 2017).

26 Bacteria do not have brains because they are unicellular, but it could be argued that bacteria have 'minds' since the receptors on the outside of their cells are designed to identify 'good' and 'bad' foods that determine in which direction they travel.

27 The clathrate-gun hypothesis is the subject of a book by Kennett and others (2002). In fairness, it should be pointed out that there are other explanations for the collapse of methane-hydrate-filled sediment piles along continental margins, including one involving changes in water temperature (associated with glacial-interglacial cycles) that simply causes the ice casings of the methane to melt, releasing the gas in large quantities almost simultaneously and leading to collapse. Scientists have worked backwards from discoveries like that of finding that collapsed material at the edge of the underwater Amazon Delta contained some 10 per cent methane hydrates, a strong indication that the dissociation of these caused the associated collapse.

28 This event is the Storegga Slide, the main part dated to 8,100 ± 250 years ago. Wave heights associated with the Storegga Slide were calculated by Stein Bondevik and colleagues (2005, *Marine and Petroleum Geology*); the scattering of stone tools consistent with tsunami impact was deduced for a site in Inverness by Alastair Dawson and colleagues (1990, *Journal of Archaeological Science*).

29 The destruction of worlds by fire, flood and earthquake is a persistent theme in Norse mythology (Gaiman 2017), perhaps most famously in anticipatory descriptions of Ragnarok, the future world-ending battle, the equivalent of Armageddon in Greco-Roman cultures.

30 How and why islands can disappear is explained in my 2009 book, *Vanished Islands and Hidden Continents of the Pacific* (Nunn 2009), now available to listen to – see https://tinyurl.com/lbrrgfa.

31   More details about the original Teonimenu story (and stories of
     other vanished islands in the vicinity) are found, together with a
     cogent geological explanation, in the work by myself and others
     (2006, *South Pacific Studies*). The transcription of an original
     story about Teonimenu by Zacchariah Haununumaesihaa'a,
     collected from Ulawa Island by Tony Heorake, is found in
     Appendix 1 of my 2009 book (Nunn 2009).

32   This remarkable story was told by Wolfe and colleagues in
     the 1994 *Journal of Geophysical Research*.

33   This extraordinary piece of cartographic detective work was
     reported by Filmer and others in the 1994 issue of *Marine
     Geophysical Researches* and is supported by indigenous stories
     recalling a 'large disaster' in the area around the year 1800.

34   Quote from p. 463 in the book on *Yurok Myths* by Kroeber
     (1976).

35   Coastal settlements throughout the Cascadia region (north-
     west USA and south-west Canada) have been affected by
     abrupt, earthquake-induced subsidence; a wealth of oral
     stories about these events is reported in the 1996 volume of
     American Antiquity.

36   Earthquakes are estimated to occur only every 2,000 years
     along the Sparta Fault. That is why the timing of the
     destructive 464 BC event may be so significant for an
     understanding of Plato's thinking about the role of
     catastrophic events in societal collapse.

37   Chapter XI of Book III of Thucydides (431 BC).
     Geoscientists remember Thucydides as the first person to
     correctly infer a connection between earthquakes and
     tsunamis. The following is a continuation of the quote used
     in the text. 'The cause, in my opinion, of this phenomenon
     must be sought in the earthquake. At the point where its
     shock has been the most violent, the sea is driven back and,
     suddenly recoiling with redoubled force, causes the
     inundation. Without an earthquake I do not see how such an
     accident could happen.'

38   Plato is generally considered the originator of the 'Socratic
     dialogue', a literary device based on oral debates. The details
     of the Atlantis story appear in two of his dialogues, *Critias*

and *Timaeus*. Plato's most prominent disciple, Aristotle, probably reflected his master's teachings when he pondered the modes of persuasion furnished by the spoken word in his *Rhetoric* (350 BC); he identified three modes, the third of which explained that persuasion is achieved through the speech itself when a truth (or an apparent truth) is proven by means of the persuasive arguments suited to the case in question. Indeed.

39   In the immediate aftermath of the explosion, one eyewitness stated that 'we saw around us a miracle, a terrible miracle. The forest was not our forest. I have never seen such a forest in my life. It was so unfamiliar. We had here a dense forest, a dark forest, an old forest. And now there was in many places no forest at all. On the mountains all the trees were lying down and it was light; one could see far away. And it was impossible to go under the mountains, through the bogs: some trees were standing there, others were down, still others were bent, and some trees had fallen one upon another. Many trees were burnt, dry trunks and moss were still burning and smoking' (part of Akulina's story, quoted in the paper by Suslov in the 2006 issue of the *Russian Institute on Anomalous Phenomena (RIAP) Bulletin*).

40   This is the English translation in the 1996 issue of *Meteoritics & Planetary Science* of part of the original account by Antenor Álvarez (1926).

41   Australian microtektites – tektites small enough to circulate some distance in the atmosphere – have been found in Chinese loess and in accumulations of eroded material around the Transantarctic Mountains.

42   During a visit to the Henbury meteorite crater field in June 1931 by a group from the Kyancutta Museum (South Australia), a journey of more than 4,800km (3,000 miles) by 'motor truck', contact was made with 'a local prospector' who supplied the Aboriginal name for this site. A neat summary of traditions by Duane Hamacher about the Henbury craters was published online in *The Conversation* on 4 March 2016 (http://theconversation.com/finding-meteorite-impacts-in-aboriginal-oral-tradition-38052).

43 One example is of Wolfe Creek crater (Western Australia),
    which formed from bolide impact about 300,000 years
    ago. Long before Western-trained scientists had worked
    out its origin, local Djaru people had stories consistent
    with this, even though its formation must have gone
    unwitnessed by people. One account by Djaru man Jack
    Jugarie states that 'a star bin fall down. It was a small star,
    not so big. It fell straight down and hit the ground. It fell
    straight down and made that hole round, a very deep hole.
    The earth shook when that star fell down' (quoted in
    Sanday 2007: 26).

44 From pp. 32–33 of *The Adnyamathanha People*, produced by
    the Education Department of South Australia.

45 This story was collected from Aboriginal informants in the
    Shoalhaven region by Ellen Anderson, between perhaps 1870
    and 1920, and comes from pp. 192–193 of *Australian Legends*
    by Charles W. Peck (1933).

46 Flood myths feature in almost every long-established culture
    (Dundes 1988), and in most cases probably represent stories of
    successive flood events.

47 A trailblazing survey of preservice (trainee) teachers in the
    United States found that 'sizeable minorities ... awaited
    more evidence' as to whether fantastic beasts (like Bigfoot
    and the Loch Ness Monster) were real (see Susan Losh,
    2011, *Journal of Science and Education*, p. 473). The study
    concluded that 'more training is needed for preservice
    educators in the critical evaluation of material evidence'.
    I agree.

48 Less charitably referred to as hobbits (which are imaginary
    creatures), the details of this extraordinary discovery –
    extraordinary because it had long been regarded as an
    orthodoxy that our species was the only one of *Homo* to have
    been around in the last 100,000 years or so – were first
    announced in the journal *Nature* on 28 October 2004 in
    three short pieces and an article.

49 See the 2016 paper in *Nature* by Sutikna and others.

50 Details are in Forth's *Beneath the Volcano* (1998).

51 See the most recent book by Forth (2008).

52 If stories of small people coexisting with larger ones (like us)
   intrigue you, I recommend the riveting and scientifically
   rigorous book *Pygmonia* (McAllister 2010).
53 This quote comes from an unpublished 1845 manuscript
   without page numbers entitled *An account of some fossil bones
   found in Darling Downs* by F. N. Isaac.
54 Quotes from Barrett (1946: 29–30).
55 The first quote in this sentence is from Opit (2001: 45),
   whose account includes many other accounts of possible or
   claimed *bunyip* sightings. The second quote, together with
   that in the following sentence, comes from the description of
   Palorchestes azael in the superbly illustrated book by Archer
   and others (1991: 115). Several scientists have suggested that
   Aboriginal stories about *bunyip* and similar creatures might be
   based on observations thousands of years ago of
   diprotodontoids like *P. azael*.
56 I am troubled by the oft-used argument that because
   Palorchestes azael became extinct some 40,000 years ago
   (which it perhaps did not – see the next note) it could not
   possibly be represented in rock art, especially given the
   compelling arguments to the contrary (see, for example, the
   paper by Murray in the 1984 issue of *Archaeology in Oceania)*.
57 The Riversleigh terrace date of 23,900 (+ 4,100 or - 2,700)
   years ago was reported in the 1997 *Memoirs of the Queensland
   Museum* by Davis and Archer. The Spring Creek date of
   19,800 ± 390 years ago is from cave deposits containing
   megafaunal remains including those of *Palorchestes azael* (see
   Tim Flannery's paper in the 1984 issue of *The Australian
   Zoologist)*. Popular accounts of this creature often claim that it
   survived until 11,000 years ago in Australia, but I have been
   unable to find a reliable source for this claim.
58 It is easy to forget that because Australia is so vast and most of
   it is so sparsely populated there is a greater potential here,
   compared to most other places, for animals regarded as
   extinct (at least recently) to be discovered alive. For example,
   science regards the Tasmanian Tiger (*Thylacinus cynocephalus*),
   an endemic thylacine, to have become extinct in the 1930s,
   yet in March 2017, in response to increasing numbers of

plausible sightings in far north Queensland, scientists began to look for it there (www.jcu.edu.au/news/releases/2017/march/fnq-search-for-the-tasmanian-tiger, accessed in March 2017). Naturally, they are keeping coy about precisely where they lay their camera traps.

59 Just to note that the American cheetah was probably slower than its modern counterpart, better at climbing trees than running. And that the camels, notwithstanding genomic evidence showing their links to African and Asian camels, appear to have been more akin in appearance and habit to modern llamas than modern camels. And that the American horse, once almost ubiquitous in ice-free North America, became extinct about 11,000 years ago – modern horses (including those in North America) are all descended from those that have existed for millions of years in Africa and Eurasia. Why one group became extinct while the other did not remains a puzzle.

60 The Antarctic Cold Reversal occurred between 12,700 and 14,400 thousand years ago (overlapping the Younger Dryas in the northern hemisphere) and was a significant departure from the long-term postglacial warming trend in which it is embedded. Patagonian megafaunal extinction took a surprisingly short time – no more than 300 years around 12,280 years ago.

61 The palaeontologist was Richard Owen, and it was in 1859 that he named Australia's marsupial lion *Thylacoleo carnifex*. Since this lion was manifestly a marsupial mammal, scientific prejudices about all such diprotodonts being herbivores led to considerable argument about whether Owen's characterisation of this beast as a carnivore was correct. Later research proved him right.

62 This study was reported by Trueman and colleagues in the 2005 issue of the *Proceedings of the National Academy of Sciences of the United States of America*.

63 Utilising amino acid racemization (AAR) and electron spin resonance (ESR) dating techniques, the revision of the Tasmanian megafaunal extinction ages suggests that humans were in fact the main cause of this (see work by Chris Turney

and others in the 2008 issue of the *Proceedings of the National Academy of Sciences of the United States of America*). Another intriguing study, covering all of Australia, reached the same conclusion. At several places around Australia, these researchers analysed sediments from both the ocean floor and onshore cave fills to determine the amount of the dung fungus, *Sporormiella*, in layers of different ages. *Sporormiella* is a proxy for herbivore biomass – the more there is, the more large-bodied herbivores (including megafauna) were living in a particular place at a particular time. It was found that *Sporormiella* levels were high between 45,000 and 150,000 years ago, but that a 'marked decline' (p. 1) occurred between 43,100 and 45,000 years ago, which is interpreted as bracketing in time the main period of megafaunal extinction (see van der Kaars, 2017, *Nature Communications*).

64 The quotes in this paragraph are from the account of how *mihirung paringmal* (thunder birds) were once hunted; it comes from pp. 92–93 of the book by Dawson (1881), who gives the phonetic names for these birds as *meeheeruung parrinmall*. I have paraphrased Dawson's account, which also dates this tradition to the time 'long ago when the volcanic hills were in a state of eruption' (p. 92), something that dates this account of the *mihirung* to perhaps 5,000–10,000 years ago.

65 The identification of *Genyornis* bones by Aboriginal people as *mihirung* was reported from excavations at Lancefield Swamp, New South Wales (Murray and Vickers-Rich 2004). The painting of a megafaunal bird, possibly *Genyornis newtoni*, at the Nawarla Gabarnmang site in northern Australia was reported and discussed by Ben Gunn and others (2011, *Australian Archaeology*) and is shown in the colour plate section.

66 The view that *Genyornis newtoni* became extinct around 47,000 years ago contrasts with the view that it may have co-existed with humans in Australia until much more recent times (Gerritsen 2011). The latter study quotes from an 1845 report by George Adney in western Victoria about the time he unearthed some giant bones near Lake Colongulac that the local (Aboriginal) Tjakut speakers stated were the remains of 'a fearful monster ... on two legs with a neck and head like

a large emeu [emu] and a breast covered with shaggy hair and killing men by hugging them with his large flappers' (p. 64), plausibly an account of the *mihirung* or *Genyornis newtoni*.

67 Quote from p. 147 of the paper by Grellet-Tinner and others in the 2016 issue of *Quaternary Science Reviews*.

68 A good article on the subject of this case of mistaken identity was published online in *The Conversation* on 14 January 2016 (http://theconversation.com/a-case-of-mistaken-identity-for-australias-extinct-big-bird-52856).

Chapter 7: Have We Underestimated Ourselves?

1 Research in Kerala showed that the local fishing community trusted its traditional knowledge of coastal risk, especially storm surges and tsunamis, more than the understanding implicit in the strategies imposed by central government (Santha, 2014, *Natural Resources Forum*). The framing of tsunami precursors in myths and other types of traditional oral knowledge saved numerous lives during tsunami events in Simeulue during the great Indian Ocean Tsunami of 2004 and in southern Pentecost during a 1999 event (McAdoo, 2006, *Earthquake Spectra*).

2 Motifs are 'the smallest element in a tale having the power to persist in tradition' (Thompson 1977: 415). Motifs define 'tale types', over 2,000 of which have been codified in the Aarne-Thompson-Uther (ATU) Index.

3 The research on the *Little Red Riding Hood* tale types was reported by Jamie Tehrani (2013, *PLOS One*).

4 Many thinkers have argued along such lines before now. To René Girard, most myths had an empirical basis, distorted representations of real events that non-literate people encoded and passed on for posterity (Golsan 2002). To Walter Ong, the replacement of vibrant spontaneous orality with colourless literacy was to be lamented; he cautioned that the world of oral knowledge was in danger of being consigned to the rubbish bin of history (Ong 1982).

5 In his magisterial *Eden in the East* (1998), Stephen Oppenheimer proposed that memories of the drowning of the Sunda Shelf (where island South-east Asia lies today), perhaps soon after the end of the Last Glacial Maximum

about 18,000 years ago, may be preserved in stories of lands
being 'pulled up' that are found in cultures all around it.

6   The earliest written accounts of Australian Aboriginal
watercraft described them as 'rafts and canoes made of logs or
sewn bark, bark or reed bundles ... the uses of these watercraft
were restricted historically to be used close to shore or to have
been restricted to visiting islands only as far as 25 km and
mostly less than 10 km offshore' (Jane Balme, 2013, *Quaternary
International*, p. 71).

Further Reading

Chapter 1: Recalling the Past

Atwater, B. F. *et al.* 2005. *The Orphan Tsunami of 1700 – Japanese Clues to a Parent Earthquake in North America.* Seattle: University of Washington Press.

Barber, E. W. & P. T. Barber. 2004. *When They Severed Earth from Sky: How the Human Mind Shapes Myth.* Princeton: Princeton University Press.

Beaglehole, E. & P. Beaglehole. 1938. *Ethnology of Pukapuka.* Vol. 150, *Bulletin.* Honolulu: BP Bishop Museum.

Bronowski, J. 1974. *The Ascent of Man.* Boston: Little, Brown.

Clark, E. 1953. *Indian Legends of the Pacific Northwest.* Berkeley: University of California Press.

Kelly, L. 2016. *The Memory Code.* Sydney: Allen and Unwin.

Kingdon, J. 1996. *Self-made Man: Human Evolution From Eden to Extinction.* Chichester: Wiley.

Ong, W. 1982. *Orality and Literacy: The Technologizing of the Word.* London: Routledge.

Piccardi, L. & W. B. Masse, eds. 2007. *Myth and Geology.* London: Geological Society of London.

Ricci, E. 1969. *I Peligni Superequani la Sicinnide e le Origini di Secinaro.* Sulmona: Italia Editoriale.

van den Berg, R. 2002. *Nyoongar People of Australia: Perspectives on Racism and Multiculturalism.* Leiden: Brill.

Vitaliano, D. 1973. *Legends of the Earth: Their Geologic Origins.* Bloomington: Indiana University Press.

Chapter 2: Words that Matter in a Harsh Land

Berndt, R. M. & C. H. Berndt. 1996. *The World of the First Australians. Aboriginal Traditional Life: Past and Present.* Canberra: Aboriginal Studies Press.

Blewett, R. S., ed. 2012. *Shaping a Nation: A Geology of Australia.* Canberra: Geoscience Australia and ANU E Press.

Carnegie, D. W. 1898. *Spinifex and Sand.* London: Arthur Pearson.

Cook, J. 1893. *Captain Cook's Journal During his First Voyage Round the World Made In H.M. Bark 'Endeavour' 1768–71 (A Literal Transcription of the Original MSS).* London: Elliot Stock.

Flannery, T. F. 1994. *The Future Eaters: An Ecological History of the Australasian Lands and People.* Chatswood: Reed.

Gammage, B. 2011. *The Biggest Estate on Earth: How Aborigines Made Australia.* Sydney: Allen and Unwin.

Giles, E. 1889. *Australia Twice Traversed: The Romance of Exploration, Being A Narrative Compiled from the Journals of Five Exploring Expeditions Into and Through Central South Australia, and Western Australia, from 1872 to 1876.* London: Sampson Low.

Gill, A. M. *et al.*, eds. 1981. *Fire and the Australia Biota.* Canberra: Australian Academy of Science.

Haygarth, H. W. 1861. *Recollections of Bush Life in Australia.* London: John Murray.

Meeham, B. & N. White, eds. 1990. *Hunter-Gatherer Demography, Past and Present.* Sydney: University of Sydney.

Mitchell, T. 1848. *Journal of an Expedition Into the Interior of Tropical Australia In Search of a Route from Sydney to the Gulf of Carpentaria.* London: Longman, Brown, Green.

Mulvaney, D. J. & J. Kamminga. 1999. *Prehistory of Australia.* Crows Nest, New South Wales: Allen and Unwin.

Parkinson, S. 1773. *A Journal of a Voyage to the South Seas, in His Majesty's Ship the Endeavour.* London: Printed privately for Stansfield Parkinson.

Sturt, C. 1834. *Two Expeditions Into the Interior of Southern Australia During the Years 1828, 1829, 1830, and 1831.* 2nd edn. Vol. 1. London: Smith, Elder.

Thomson, D. 1975. *Bindibu Country.* Melbourne: Thomas Nelson.

van Gennep, A. 1906. *Mythes et Légendes d'Australie.* Paris: Guilmoto.

Chapter 3: Australian Aboriginal Memories of Coastal Drowning

Berndt, C. H. & R. M. Berndt. 1994. *The Speaking Land: Myth and Story in Aboriginal Australia.* Rochester, Vermont: Inner Traditions/Bear & Co.

Berndt, R. M. & C. H. Berndt. 1996. *The World of the First Australians. Aboriginal Traditional Life: Past and Present.* Canberra: Aboriginal Studies Press.

Cane, S. 2002. *Pila Nguru: The Spinifex People.* North Fremantle, Western Australia: Fremantle Arts Centre Press.

Cooper, H. M. 1955. *The Unknown Coast: A Supplement.* Adelaide: Advertiser Printing Office.

Dawson, J. 1881. *Australian Aborigines: The Languages and Customs of Several Tribes of Aborigines in the Western District of Victoria, Australia.* Melbourne: George Robertson.

Dixon, R. M. 1972. *The Dyirbal Language of North Queensland.* Vol. 40. Cambridge: Cambridge University Press.

Dixon, R. M. 1980. *The Languages of Australia.* Cambridge: Cambridge University Press.

Dixon, R. M. W. 1977. *A Grammar of Yidiny, Cambridge Studies in Linguistics.* Cambridge: Cambridge University Press.

Dixon, R. M. W. 1991. *Words of Our Country: Stories, Place Names and Vocabulary in Yidiny, the Aboriginal Language of the Cairns-Yarrabah Region.* St Lucia: University of Queensland Press.

Falkiner, S. & A. Oldfield. 2000. *Lizard Island: The Journal of Mary Watson.* Crows Nest, New South Wales: Allen and Unwin.

Fison, L. & A. W. Howitt. 1880. *Kamilaroi and Kurnai.* Melbourne: George Robertson.

Flinders, M. 1814. *A Voyage to Terra Australis; undertaken for the purpose of completing the discovery of that vast country, and prosecuted in the years 1801, 1802, and 1803, in His Majesty's ship The Investigator, and subsequently in the armed vessel Porpoise and Cumberland Schooner.* 2 vols. London: G and W Nicol.

Flood, J. 2006. *The Original Australians: Story of the Aboriginal People*. Crows Nest, New South Wales: Allen and Unwin.

Gribble, E. R. B. 1932. *The Australian Aboriginal*. Sydney: Angus and Robertson.

Hiatt, L. R. 1978. *Australian Aboriginal Concepts*. Canberra: Australian Institute of Aboriginal Studies.

Isaacs, J. 1980. *Australian Dreaming: 40,000 Years of Aboriginal History*. Sydney: Lansdowne Press.

Krichauff, S. 2011. *Nharrungga Wargunni Bugi-Buggillu: A Journey Through Narungga History*. Kent Town, South Australia: Wakefield Press.

Kuhn, D. & C. Freeman, eds. 2012. *Georges River Estuary Handbook*. Georges River, New South Wales: Georges River Combined Councils Committee.

Lampert, R. J. 1981. *The Great Kartan Mystery*. Vol. 5. *Terra Australis*. Canberra: Australian National University.

Massola, A. 1968. *Bunjil's Cave: Myths, Legends and Superstitions of the Aborigines of South-east Australia*. Melbourne: Lansdowne.

McCrae, H., ed. 1934. *Georgiana's Journal: Melbourne a Hundred Years Ago [Diary of Georgiana McCrae]*. Sydney: Angus and Robertson.

Meyer, H. E. A. 1846. *Manners and Customs of the Aborigines of the Encounter Bay Tribes, South Australia*. Adelaide: Dehane.

Moore, G. F. 1884. *Diary of Ten Years Eventful Life of an Early Settler in Western Australia; and also A Descriptive Vocabulary of the Language of the Aborigines*. London: Walbrook (facsimile edition 1978 by University of Western Australia Press).

Morris, J. 2001. *The Tiwi: From Isolation to Cultural Change*. Darwin: NTU Press.

Mulvaney, J. & N. Green. 1992. *Commandant of Solitude: The Journals of Captain Collet Barker, 1828–1831*. Melbourne: Melbourne University Press at the Miegunyah Press.

Niall, B. 1994. *Georgiana: A Biography of Georgiana McCrae, Painter, Diarist, Pioneer*. Melbourne: Melbourne University Press.

Noonuccal, O. 1990. *Australian Legends and Landscapes.* Sydney: Random House.

Nunn, P. D. 2007. *Climate, Environment and Society in the Pacific During the Last Millennium.* Amsterdam: Elsevier.

Nunn, P. D. 2009. *Vanished Islands and Hidden Continents of the Pacific.* Honolulu: University of Hawai'i Press.

Osborne, C. R. 1974. *The Tiwi Language.* Canberra: Australian Institute of Aboriginal Studies (Australian Aboriginal Series 55, Linguistic Series 21).

Parker, K. L. 1959. *Australian Legendary Tales.* Sydney: Angus and Robertson.

Piccardi, L. & W. B. Masse, eds. 2007. *Myth and Geology.* London: Geological Society of London.

Reed, A. W. 1965. *Myths and Legends of Australia.* Sydney: Reed.

Reed, A. W. 1993. *Aboriginal Myths, Legends and Fables.* Chatswood, New South Wales: Reed.

Roberts, A. & C. P. Mountford. 1989. *The Dawn of Time: Australian Aboriginal Myths.* Blackwood, South Australia: Art Australia.

Robinson, G. A. 2008. *Friendly Mission: The Tasmanian Journals and Papers of George Augustus Robinson 1829–1834.* 2nd edn. Launceston: Queen Victoria Museum and Art Gallery and Quintus Publishing.

Rogers, H. 1957. *The Early History of the Mornington Peninsula.* Mornington, Melbourne: Mornington Leader.

Roughsey, D. 1971. *Moon and Rainbow: The Autobiography of an Aboriginal.* Sydney: Reed.

Russell, A. 1934. *A Tramp-Royal in Wild Australia.* London: Cape.

Ryan, L. 2012. *Tasmanian Aborigines: A History Since 1803.* Sydney: Allen and Unwin.

Smith, J. 1880. *Booandik Tribe of South Australian Aborigines: A Sketch of their Habits, Customs, Legends and Language.* Adelaide: Government Printer.

Smith, W. R. 1930. *Myths and Legends of the Australian Aboriginals.* London: Harrap.

Taplin, G. 1873. *The Narrinyeri: An Account of the Tribes of South Australian Aborigines Inhabiting the Country Around the Lakes Alexandrina, Albert and Coorong, and the Lower Part of the River Murray.* Adelaide: Government Printer.

Tench, W. 1793. *A Complete Account of the Settlement at Port Jackson, in New South Wales, Including an Accurate Description of the Situation of the Colony; of the Natives; and of its Natural Productions.* London: Nicol and Sewell.

Unaipon, D. 2001. *Legendary Tales of the Australian Aborigines.* Melbourne: Miegunyah Press.

Welsby, T. 1967. *The Collected Works of Thomas Welsby.* A. K. Thomson (ed.). 2 vols. Brisbane: Jacaranda Press.

Wright, R. V. S. 1971. *Archaeology of the Gallus Site, Koonalda Cave.* Canberra: Australian Institute of Aboriginal Studies.

Chapter 4: The Changing Ocean Surface

Britton, H. 1870. *Fiji in 1870, Being the Letters of 'The Argus' Special Correspondent.* Melbourne: Samuel Mullen.

Dixon, R. M. 1980. *The Languages of Australia.* Cambridge: Cambridge University Press.

Gaffney, V. *et al.* 2009. *Europe's Lost World: The Rediscovery of Doggerland.* York: Council for British Archaeology.

Hopley, D. *et al.* 2007. *Geomorphology of the Great Barrier Reef: Development, Diversity and Change.* Cambridge: Cambridge University Press.

Mol, D. *et al.* 2008. *The Saber-toothed Cat.* Norg: Uitgeverij DrukWare.

Nunn, P. D. 1999. *Environmental Change in the Pacific Basin: Chronologies, Causes, Consequences.* New York: Wiley.

Nunn, P. D. 2007. *Climate, Environment and Society in the Pacific During the Last Millennium.* Amsterdam: Elsevier.

Chapter 5: Other Oral Archives of Ancient Coastal Drowning

Ashbee, P. 1974. *Ancient Scilly: From the First Farmers to the Early Christians.* Newton Abbot: David and Charles.

Barber, E. W. & P. T. Barber. 2004. *When They Severed Earth from Sky: How the Human Mind Shapes Myth*. Princeton: Princeton University Press.

Borlase, W. 1758. *The Natural History of Cornwall*. Oxford: Oxford University Press.

Camden, W. 1590. *Britannia siue Florentissimorum regnorum, Angliæ, Scotiæ, Hiberniæ, et insularum adiacentium ex intima antiquitate chorographica descriptio*. London: Eliot's Court Press.

Carew, R. 1723. *The Survey of Cornwall. And an Epistle Concerning the Excellencies of the English Tongue*. London: Samuel Chapman.

Doble, G. H. 1962. *The Saints of Cornwall, Part II*. Truro: Dean and Chapter of Truro Cathedral.

Flemming, N. 1972. *Cities in the Sea*. 2nd edn. London: New England Library.

Fox, C. & B. Dickens, eds. 1950. *The Early Cultures of Northwest Europe*. Cambridge: Cambridge University Press.

Guyot, C. 1979. *The Legend of the City of Ys*. Translated by D. Cavanagh. Amherst: University of Massachusetts Press.

Jones, G. & T. Jones. 2001. *The Mabinogion*. New York: Knopf.

North, F. J. 1957. *Sunken Cities: Some Legends of the Coast and Lakes of Wales*. Cardiff. University of Wales Press.

Nunn, P. D. 2009. *Vanished Islands and Hidden Continents of the Pacific*. Honolulu: University of Hawai'i Press.

Poingdestre, J. 1889. *Caesarea; or, A Discourse of the Island of Jersey*. St Helier: Le Feuvre.

Ramaswamy, S. 2004. *The Lost Land of Lemuria: Fabulous Geographies, Catastrophic Histories*. Berkeley: University of California Press.

Rao, S. R. 1999. *The Lost City of Dvaraka*. New Delhi: Aditya Prakashan.

Sébillot, P. 1899. *Légendes Locales de la Haute-Bretagne*. Nantes: Société des Bibliophiles Bretons.

Stanley, J.-D. 2007. *Geoarchaeology: Underwater Archaeology in the Canopic Region in Egypt*. Oxford: Oxford Centre for Maritime Archaeology.

Stewart, I. S. & C. Vita–Finzi, eds. 1998. *Coastal Tectonics*. London: Geological Society of London.

Wilson, J. M. 1870. *The Imperial Gazetteer of England and Wales*. Vol. 1. Edinburgh: Fullarton.

Chapter 6: What Else Might We Not Realise We Remember?

Alvarez, A. 1926. *El Meteorito del Chaco*. Buenos Aires: Peuser.

Archer, M. *et al.* 1991. *Riversleigh, The Story of Animals in Ancient Rainforests of Inland Australia*. Sydney: Reed Books.

Barrett, C. 1946. *The Bunyip and Other Mythical Monsters and Legends*. Melbourne: Reed and Harris.

Blong, R. 1982. *The Time of Darkness: Local Legends and Volcanic Reality in Papua New Guinea*. Canberra: Australian National University Press.

Charachidzé, G. 1986. *Prométhée ou le Caucase*. Paris: Flammarion.

Dawson, J. 1881. *Australian Aborigines: The Languages and Customs of Several Tribes of Aborigines in the Western District of Victoria, Australia*. Melbourne: George Robertson.

Dixon, R. 1984. *Searching for Aboriginal Languages: Memoirs of a Field Worker*. Brisbane: University of Queensland Press.

Dixon, R. M. W. 1991. *Words of Our Country: Stories, Place Names and Vocabulary in Yidiny, the Aboriginal Language of the Cairns-Yarrabah Region*. St Lucia: University of Queensland Press.

Dundes, A., ed. 1988. *The Flood Myth*. Berkeley: University of California Press.

Forth, G. 1998. *Beneath the Volcano: Religion, Cosmology and Spirit Classification Among the Nage of Eastern Indonesia*. Leiden: KITLV Press.

Forth, G. 2008. *Images of the Wildman in Southeast Asia: An Anthropological Perspective*. Oxford: Routledge.

Gaiman, N. 2017. *Norse Mythology*. London: Bloomsbury.

Gerritsen, R. 2011. *Beyond the Frontier: Explorations in Ethnohistory*. Canberra: Batavia Online Publishing.

Howitt, A. W. 1904. *The Native Tribes of South-east Australia.* London: Macmillan.

Johnson, R. W. 2013. *Fire Mountains of the Islands: A History of Volcanic Eruptions and Disaster Management in Papua New Guinea and the Solomon Islands.* Canberra: ANU E Press.

Kennett, J. P. *et al.* 2002. *Methane Hydrates in Quaternary Climate Change: The Clathrate Gun Hypothesis.* Washington DC: American Geophysical Union.

Kroeber, A. L. 1976. *Yurok Myths.* Berkeley: University of California Press.

McAllister, P. 2010. *Pygmonia: My Quest for the Secret Land of the Pygmies.* Brisbane: University of Queensland Press.

McCalla, A. 2013. *The Creationist Debate: The Encounter Between the Bible and the Historical Mind.* 2nd edn. New York: Bloomsbury.

Mjöberg, E. 1918. *Bland Stenåldersmänniskor i Queenslands Wildmarker (Amongst Stone Age People in the Queensland Wilderness).* Trans. S. M. Fryer. Stockholm: Albert Bonniers.

Murray, P. F. & P. Vickers-Rich. 2004. *Magnificent Mihirungs: The Colossal Flightless Birds of the Australian Dreamtime.* Bloomington: Indiana University Press.

Nunn, P. D. 1994. *Oceanic Islands.* Oxford: Blackwell.

Nunn, P. D. 2009. *Vanished Islands and Hidden Continents of the Pacific.* Honolulu: University of Hawai'i Press.

Opit, G. 2001. 'The Bunyip'. *Myths and Monsters 2001*, Sydney.

Peck, C. W. 1933. *Australian Legends.* Melbourne: Lothian.

Sanday, P. R. 2007. *Aboriginal Paintings of the Wolfe Creek Crater: Track of the Rainbow Serpent.* Philadelphia: University of Pennsylvania Press.

Smith, J. 1880. *Booandik Tribe of South Australian Aborigines: A Sketch of Their Habits, Customs, Legends and Language.* Adelaide: Government Printer.

Smyth, R. B. 1878. *The Aborigines of Victoria: With Notes Relating to the Habits of the Natives of Other Parts of Australia and Tasmania.* London: John Ferres.

Vitaliano, D. 1973. *Legends of the Earth: Their Geologic Origins.* Bloomington, Indiana: Indiana University Press.

Winchester, S. 2003. *Krakatoa: The Day the World Exploded: August 27, 1883.* London: Viking.

Chapter 7: Have We Underestimated Ourselves?

Golsan, R. J. 2002. *René Girard and Myth: An Introduction.* New York: Routledge.

Ong, W. 1982. *Orality and Literacy: The Technologizing of the Word.* London: Routledge.

Oppenheimer, S. 1998. *Eden in the East: The Drowned Continent of Southeast Asia.* London: Weidenfeld and Nicolson.

Thompson, S. 1977. *The Folktale.* Oakland: University of California Press.

Acknowledgements

For guiding me through ways of hearing, reading and understanding ancient stories, I would especially like to thank Marjorie Le Berre, Earle de Blonville, Jen Carter, Rita Compatangelo-Soussignan, Axel Creach, Marie-Yvane Daire, Robert Dixon, Paul Geraghty, Diane Goodwillie, Duane Hamacher, Roselyn Kumar, Frédéric Le Blay, Bruce Masse, Sepeti Matararaba, Elia Nakoro, Petra Nunn, Nick Reid, John Runman, Margaret Sharpe and Dorothy Vitaliano as well as numerous others who have encouraged and challenged me. I thank the competent team at Bloomsbury, especially Jim Martin and Anna MacDiarmid, as well as my copy editor Krystyna Mayer. My family could not have been more supportive – to them this book is dedicated.

Index